01-CTS-1

UNDERSTANDING FACTORS AFFECTING POLYMER DEMAND FOR CONDITIONING AND DEWATERING

by:

Matthew J. Higgins, Ph.D. and
Yen-Chih Chen, Ph.D.
Department of Civil and Environmental Engineering
Bucknell University

Sudhir N. Murthy, Ph.D., P.E.
District of Columbia Water and Sewer Authority

2006

WERF

The Water Environment Research Foundation, a not-for-profit organization, funds and manages water quality research for its subscribers through a diverse public-private partnership between municipal utilities, corporations, academia, industry, and the federal government. WERF subscribers include municipal and regional water and wastewater utilities, industrial corporations, environmental engineering firms, and others that share a commitment to cost-effective water quality solutions. WERF is dedicated to advancing science and technology addressing water quality issues as they impact water resources, the atmosphere, the lands, and quality of life.

For more information, contact:
Water Environment Research Foundation
635 Slaters Lane, Suite 300
Alexandria, VA 22314-1177
Tel: (703) 684-2470 Fax: (703) 299-0742
www.werf.org werf@werf.org

This report was co-published by the following organization. For nonsubscriber sales information, contact:

IWA Publishing
Alliance House, 12 Caxton Street
London SW1H 0QS, United Kingdom
Tel: +44 (0) 20 7654 5500
Fax: +44 (0) 20 7654 5555
www.iwapublishing.com
publications@iwap.co.uk

Library of Congress Catalog Card Number: 2004108839

IWAP ISBN: 978-1-84339-726-7

Bucknell University, District of Columbia Water and Sewer Authority

ACKNOWLEDGMENTS

The Project Team would like to thank WERF for their support of the research. In addition, we would like to thank the many personnel at plants that helped us with sampling and provided numerous hours of in-kind contributions to this project. Finally, the Project Team thanks the Project Subcommittee (listed below) for their valuable guidance and oversight of the project and the WERF Project Managers, Mary Strawn and Amit Pramanik, for their support and assistance during this project.

Report Preparation

Principal Investigators:
Matthew J. Higgins, Ph.D.
Bucknell University

Sudhir N. Murthy, Ph.D., P.E.
District of Columbia Water & Sewer Authority (DCWASA)

Yen-Chih Chen, Ph.D.
Bucknell University

Project Subcommittee
John T. Novak, Ph.D., P.E. Chair
Virginia Tech

Mohammad Abu-Orf, Ph.D.
Siemens Water Technologies

Don Hahn, P.E.
Procter and Gamble (Retired)

David Lordi, Ph.D.
Metropolitan Water Reclamation District of Greater Chicago

Ernest Mayer, Ph.D.
DuPont Engineering

Ronald B. Sieger, P.E., DEE
CH2M Hill

P. Aarne Vesilind, Ph.D., P.E.
Bucknell University

Water Enviroment Research Foundation Staff

Director of Research:	Daniel M. Woltering, Ph.D.
Senior Program Director:	Amit Pramanik, Ph.D.
Project Manager:	Mary Strawn

ABSTRACT AND BENEFITS

Abstract:

Polymer demand for sludge conditioning varies considerably for different digestion processes as well as the same digestion processes at different locations and the reasons for these differences are not known. The objectives of this research were to develop a mechanistic understanding of the factors contributing to differences in polymer demand. The differences were hypothesized to depend on both the amount of biocolloidal matter in a solids sample and the characteristics of the conditioning polymer. The first phase of the research examined which component of the solids creates the polymer demand and the second phase focused on understanding the interactions of polymers with the components of solids. The samples collected and analyzed included MLSS, RAS, conventional aerobically and anaerobically digested, thermophilic anaerobically digested, temperature phased anaerobically digested, and an auto-thermal aerobically digested biosolids. Results from the first phase showed a good linear correlation between the optimum polymer dose (OPD) and biocolloid concentration as measured by soluble protein and polysaccharide concentration. This result suggested that small particles (less than a mean size of 4.2 μm) created a significant portion of the polymer demand, especially for digested biosolids. Conditioning samples with different polymers (with varying charge density, molecular weight, and structure) demonstrated that polymer charge density and molecular weight were the key characteristics in determining OPD. The shear associated with several full-scale dewatering devices was measured and quantified using the unitless Gt term. The data from this research were used to develop an empirical model to estimate polymer demand given solids characteristics, polymer characteristics, and the shear associated with dewatering. The model was applied to a full-scale trial to estimate polymer dosage. In addition, the concepts were applied to suggest mechanistic explanations for methods to reduce polymer demand (such as ferric chloride addition, polymer addition location, and centrifuge torque).

Benefits

- Demonstrates that biocolloids, or particles less than 4.2 μm which are made up primarily of protein and polysaccharides, create a significant amount of the polymer demand for digested biosolids.
- Examines the impact of polymer characteristics such as charge density, molecular weight, and structure on polymer demand.
- Characterizes the shear sensitivity for sludges from different digestion processes (such as thermophilic, mesophilic, anaerobic and aerobic digestion) as well as undigested sludges.
- Develops the first model to predict polymer demand given the digestion type, polymer characteristics, and dewatering device.
- Demonstrates and describes techniques that can be used to reduce polymer demand for conditioning and dewatering, such as addition of iron to reduce biocolloid concentration.

Keywords: biopolymer, cationic polymer, shear, biosolids

TABLE OF CONTENTS

Acknowledgments iii
Abstract and Benefits iv
List of Tables ivii
List of Figures iviii
List of Acronyms and Abbreviations xi
Executive Summary ES-1

1.0 Background and Objectives 1-1
- 1.1 Introduction 1-1
- 1.2 Hypothesis for Interaction Between Polymer and Bioflocs (Biocolloid Theory) 1-2
- 1.3 Summary of Research Needs 1-3
- 1.4 Research Objectives 1-4

2.0 Methods and Materials 2-1
- 2.1 Experimental Approach 2-1
- 2.2 Analytical Approach and Methods 2-1

3.0 Sludge Properties Impacting Polymer Demand 3-1
- 3.1 Objective I – Verifying the Biocolloid Theory - Polymer I 3-1
  - 3.1.1 Overview of Study 3-1
  - 3.1.2 Role of Protein 3-2
  - 3.1.3 Role of Polysaccharides 3-5
  - 3.1.4 SCOD as a Surrogate Measure for Biocolloids 3-8
- 3.2 Impact of Other Sludge Characteristics and Process Parameters 3-10
  - 3.2.1 Role of Metals. 3-10
  - 3.2.2 Effect of Process Parameters. 3-14
- 3.3 Objective I - Verifying the Biocolloid Theory – Effect of Biocolloid Size Fractions 3-14
- 3.4 Conclusions and Implications 3-18
  - 3.4.1 Conclusions 3-18
  - 3.4.2 Implications 3-18

4.0 Effect of Polymer Characteristics on Polymer Demand 4-1
- 4.1 Effect of Polymer Charge and Configuration 4-1
- 4.2 Effect of Polymer Molecular Weight 4-8
- 4.3 Conclusions and Implications 4-11
  - 4.3.1 Conclusions 4-11
  - 4.3.2 Implications 4-12

5.0 Impact of Shear on Polymer Demand 5-1
- 5.1 Calibration of Mixer 5-1
- 5.2 Impact of Shear on Polymer Demand 5-1
- 5.3 Comparison of Different Sludge Types Response to Shear 5-4
- 5.4 Impact of Different G Values on Shear Response 5-6

5.5 Evaluation of Polymer Characteristics and Their Response to Shear. 5-9
5.5.1 Effect of Molecular Weight 5-9
5.5.2 Effect of Polymer Charge and Structure. 5-10
5.6 Other Impacts of Shear – Cake Solids and Cake Odors 5-13
5.6.1 Effect of Shear and Polymer Type on Dewatering Properties and Cake Solids Content 5-13
5.6.2 Impact of Shear on Odor Production from Biosolids Cake 5-18
5.7 Conclusions and Implications 5-21
5.7.1 Conclusions 5-21
5.7.2 Implications 5-21

6.0 GT of Full-Scale Dewatering Equipment 6-1
6.1 Background 6-1
6.2 Gt of Full-Scale Dewatering Equipment 6-1
6.3 Conclusions and Implications 6-3
6.3.1 Conclusions 6-3
6.3.2 Implications 6-3

7.0 Research Applications:1Can We Predict Polymer Demand? 7-1
7.1 Development of a Universal Predictive Model 7-1
7.2 Comparison of Universal Model with Field Data 7-3
7.3 Model Applications 7-6
7.4 Conclusions and Implications 7-6

8.0 Applications of Research II: Reducing Polymer Demand 8-1
8.1 Overview 8-1
8.2 Reducing Biocolloid Concentration 8-1
8.2.1 Role of Digestion 8-1
8.2.2 Chemicals and Amendments 8-2
8.2.3 Laboratory Experiments with FeCl3 Addition 8-3
8.2.4 Full-Scale Study with FeCl3 Addition 8-6
8.2.5 Conclusions for Fe Addition 8-7
8.3 Reducing Shear 8-7
8.3.1 Effect of Polymer Addition Point 8-7
8.3.2 Effect of Centrifuge Torque 8-9
8.3.3 Conclusions for Polymer Addition Location and Torque Studies 8-9
8.4. Polymer Characteristics and Dosing 8-9

9.0 Summary and Additional Research Needs 9-1
9.1 Overview 9-1
9.2 Future Research Needs 9-1

References R-1

LIST OF TABLES

3-1 Summary of Results from Batch Sampling Tests 3-2
3-2 Summary of Linear Correlation Between OPD and Cations 3-13
3-3 Correlations Between Different Process Parameters 3-14
3-4 Summary of Results from Batch Tests Examining Different Biopolymer Size Fractions (Data Sorted by Optimum Polymer Dose) 3-16
4-1 Summary of Different Polymer Characteristics 4-1
4-2 Summary of Polymer Characteristics Used in Conditioning Study 4-2
4-3 Summary of Polymer Characteristics for Effect of MW Study (All Cytec Polymers)... 4-8
5-1 Summary of Gt vs OPD Slopes and Intercept Data 5-6
5-2 Summary of Four Polymers Tested 5-10
6-1 Summary of Gt Values for Different Full-Scale Dewatering Devices 6-1
6-2 Comparison of Gt Values for Different Dewatering Devices Analyzed in This Study... 6-2
7-1 Regression Results for Multi-Parameter Model to Predict Polymer Demand 7-2
7-2 Summary of Input Values for Model, and Calculated OPD for the Given Conditions ... 7-5
8-1 Effect of Ferric Chloride Addition on Polymer Demand 8-6
8-2 Effect of Ferric Chloride Addition on Cake Solids and Centrate Quality 8-6
8-3 Summary of Results from Field Trial with Different Polymer Addition Locations........ 8-8
8-4 Summary of Data for Torque Setting Study 8-9

LIST OF FIGURES

1-1 Reaction 1, Showing the Unattached Biocolloids and the Floc in Solution 1-2
1-2 Reaction 2, Depiction of Unattached Biopolymer and Cationic Polymer Resulting in Coagulation. (Note: Not to Scale) 1-3
3-1 Relationship Between Optimum Polymer Dose and Soluble Protein Concentration for all Samples 3-3
3-2 Relationship Between Optimum Polymer Dose and Soluble Protein Concentration for all Samples Except Thermophilic Samples 3-3
3-3 Relationship Between Optimum Polymer Dose and Soluble Protein Concentration for Anaerobically Digested Samples Only 3-4
3-4 Relationship Between ASX and GLX and OPD for Anaerobically Digested Samples .. 3-5
3-5 Relationship Between GLX and ASX and Soluble Protein 3-5
3-6 Relationship Between OPD and Solution Protein Plus Polysaccharide Concentration for all Samples Except Thermophilic 3-6
3-7 Relationship Between OPD and Solution Protein Plus Polysaccharide Concentration for Anaerobically Digested Samples 3-6
3-8 Soluble Protein and CST as a Function of Polymer Dose 3-7
3-9 Relationship Between Solution SCOD and Soluble Protein and Polysaccharides for All the Samples 3-9
3-10 Relationship Between Solution COD and Soluble Protein and Polysaccharides for Anaerobically Digested Samples 3-10
3-11 Relationship Between OPD and the Total Fe Concentration of Mesophilic and Thermophilic Anaerobically Digested Samples 3-12
3-12 Relationship Between OPD and the Total Fe Concentration of Mesophilic Anaerobically Digested Samples 3-12
3-13 Relationship Between Soluble Protein and Polysaccharide Concentration and Total Fe Concentration in Anaerobically Digested Samples 3-13
3-14 Particle Size Distribution of ATAD and Anaerobically Digested Samples 3-15
3-15 Relationship Between Different Protein and Polysaccharide Size Fractions and OPD for Different Samples 3-16
3-16 Relationship Between Different COD Size Fractions and OPD For Different Samples. (r^2 = 0.85-0.87 for all Lines) 3-17
3-17 OPD vs 4.2 μm Size Fraction of Protein and Polysaccharide Showing Different Sample Types (A=Aerobically Digested, An=Anaerobically Digested, M=MLSS, T=TPAD, W=Waste Activated Sludge) 3-17
4-1 Optimum Polymer Dose Expressed as g/kg and meq/Kg for Polymers with Varying Charge Density 4-2
4-2 Schematic of Polymer Loops and Tails Showing Unused Charge of Polymer 4-3
4-3 Effect of Polymer Addition on Soluble Protein Concentration Remaining in Solution .. 4-4
4-4 Effect of Polymer Addition on Soluble Polysaccharide Concentration Remaining in Solution 4-4
4-5 Relationship Between OPD on a Mass Basis and Polymer Charge Density for Different Polymer Configurations (L=Linear, B=Branched, H=Hydrophobic) 4-5
4-6 Relationship Between OPD on an Equivalent Basis and Polymer Charge Density for Different Polymer Configurations (L=Linear, B=Branched, H=Hydrophobic) 4-5

4-7 Protein Plus Polysaccharide Concentration Remaining in Solution at the OPD as a Function of the Polymer Charge Density, Size Fraction of <8 μm (L=Linear, B=Branched, H=Hydrophobic) 4-6
4-8 Protein Plus Polysaccharide Concentration Remaining in Solution at the OPD as a Function of the OPD (Size Fraction Of <8 μm) 4-7
4-9 CST at the OPD as a Function of the Polymer Charge Density 4-7
4-10 OPD for Anaerobically Digested Sludge for Polymer Set with Varying Molecular Weight 4-9
4-11 OPD for Aerobically Digested Sludge for Polymer Set with Varying Molecular Weight 4-10
4-12 OPD for Industrial Waste Activated Sludge for Polymer Set with Varying Molecular Weight 4-10
4-13 SCOD and CST at the OPD for the Different MW Polymers and Anaerobically Digested Sample 4-11
5-1 Calibration Curve for Determining Velocity Gradient, G, as a Function of Mixer rpm. 5-1
5-2 Gt vs CST For Similar Polymer Dose on an Equivalent Basis (18 meq/kg DS) for Three Different Types of Polymers 5-2
5-3 Polymer Dose as a Function of Gt Values Giving the Minimum CST. 5-3
5-4 Gt of Minimum CST for Three Different Polymers at Different Dosage on an Charge Basis 5-4
5-5 Relationship Between CST and Polymer Dose for the Different Shear Experiments (H=hydrophobic, L=linear, B=branched) 5-4
5-6 Comparison of Shear Response for Four Different Digested Sludge Types Using the Same Linear Polymer. 5-5
5-7 Effect of G on OPD for Linear Polymer and Anaerobically Digested Sludge 5-7
5-8 Effect of G on OPD for Branched Polymer and Anaerobically Digested Sludge 5-7
5-9 Comparison of Linear and Branched Polymer Response to Gt Values for Anaerobically Digested Sludge 5-8
5-10 OPD vs Gt for WAS Sample Using Linear Polymer 5-8
5-11 Effect of MW on Response of Polymer to Shear During Conditioning. 5-9
5-12 Effect of Gt on OPD for Four Different Polymers for Anaerobically Digested Sample 5-11
5-13 OPD versus Gt for WAS and Linear Polymers with Varying Charge Density 5-12
5-14 OPD versus Gt for WAS and Polymers with Varying Structure. 5-12
5-15 CST at the Optimum Polymer Dose for Different Gt Values and Polymer Types 5-14
5-16 Effect of Gt on the Average Cake Solids Content from Vacuum Filtration 5-15
5-17 Effect of Gt on the Average Cake Solids Content from a Belt Filter Press Simulator. 5-16
5-18 Relationship Between BFP Cake Solids and CST at the OPD for Different Polymers. 5-17
5-19 Micrographs of Solids Samples That Were Conditioned Using Different Shear, Magnification of 100x for all Pictures 5-18
5-20 Effect of Shear on Odor Production from the Cake 5-19
5-21 Comparison of Total Volatile Organic Sulfur Compounds (TVOSC) Produced from Cakes Produced by Conditioning with Two Different Polymer Types 5-20
6-1 Comparison of Gt Values for Dewatering Equipment 6-3
7-1 Centrate TS and Cake Solids as a Function of Polymer Dose for an Experimental Full-Scale Trial 7-4
7-2 Comparison of Polymer Performance in Terms of Centrate TS for the HS Centrifuge and NE Sample 7-4

7-3 Comparison of Polymer Performance in Terms of Centrate TS for the MS Centrifuge and NE Sample 7-5
7-4 Estimated Polymer Demand for Anaerobically Digested Biosolids and Different Dewatering Equipment 7-6
8-1 CST Profile for $FeCl_3$ Addition 8-3
8-2 Effect of Ferric Chloride Dose on Polymer Dose for Conditioning 8-4
8-3 Effect of Ferric Chloride Addition on the Optimum Polymer Dose 8-5
8-4 Effect of Ferric Chloride (0.1 kg/kg) and Polymer Dose on Soluble Biopolymer and SCOD 8-5

LIST OF ACRONYMS AND ABBREVIATIONS

ATAD	Autothermal aerobic digestion
BFPS	Belt filter press simulator
CST	Capillary Suction Time
DS	Dry Solids
G	Velocity Gradient, Units of s^{-1}
HSC	High Solids Centrifuge
ICP	Inductively Coupled Plasma
meq	Milliequivalents
MLSS	Mixed Liquor Suspended Solids
MSC	Medium Solids Centrifuge
MW	Molecular Weight
OPD	Optimum Polymer Dose
PCD	Polymer Charge Density
PWD	Philadelphia Water Department
SCOD	Soluble Chemical Oxygen Demand
SRF	Specific resistance to filtration
Std. Dev.	Standard Deviation
t	Time, units of seconds, s
TPAD	Temperature Phased Anaerobic Digestion
TSS	Total Suspended Solids
VSS	Volatile Suspended Solids
WAS	Waste Activated Sludge
WLSSD	Western Lake Superior Sanitation District

EXECUTIVE SUMMARY

The objective of this project was to develop a better understanding of the factors that impact polymer demand for conditioning and dewatering of biosolids from wastewater treatment, especially the interactions of these factors with each other. The three main factors that were studied included the sludge characteristics, polymer characteristics, and dewatering equipment. The goal was to use this information to develop a model which could be applied to estimate polymer demand for a given treatment scenario, such as a certain digestion process and dewatering device. Plant operators and engineers could then use this model to understand how site-specific variables or changes in operations would impact polymer demand for conditioning and dewatering. In addition, the results could be used to develop methods to reduce polymer requirements for dewatering.

The main hypothesis for the cause of polymer demand was the concentration of biocolloidal material in solution. These biocolloids are made up of microbially derived biopolymers, mainly protein and polysaccharides. The highly hydrated nature of these biopolymers gives the biocolloid a gel-like consistency, and the negatively charged functional groups of the biopolymer contribute to the overall negative charge of the biocolloids and larger floc particles. Results from this work showed a linear correlation between the amount of biocolloids in solution and the polymer dose that was determined using the same polymer and mixing protocol. Digestion type had a large impact on the biocolloid concentration, with thermophilic digestion processes having a much greater concentration of biocolloids compared to mesophilic digestion processes, and undigested sludges typically having the lowest biocolloid content. The polymer demand was greatest for thermophilic processes, followed by mesophilic, and undigested sludges.

Polymers are manufactured that have a number of different characteristics, such as charge type and density, molecular weight, structure, and chemistry. This study focused on cationic polymers since these are the polymers primarily used in the wastewater industry. Over 40 different polymers were tested from different suppliers, and the polymer characteristics that varied included the charge density, molecular weight, structure and chemistry. For low shear dewatering, molecular weight and charge density were found to be two key characteristics for dewatering. Polymers required a molecular weight of greater than approximately 1 million to condition the sludges such that good dewatering properties were achieved. When polymer dose was normalized for the charge density, the structure and chemistry did not greatly impact the optimum polymer dose (OPD). As the polymer charge density increased, the OPD decreased when polymer dose was expressed on a mass basis (g polymer/kg of dry solids or DS), but increased when expressed on a charge basis (meq/kg DS) for lower shear dewatering. The dewatering properties were slightly better with the higher-charged polymers. Structured polymers with different degrees of branching and/or cross-linking did not appear to have significant advantages for the shear conditioning and dewatering tests used in this research compared to non-structured polymers of similar molecular weight and charge.

Using a standardized laboratory technique to measure shear, a mixer was calibrated to impart known levels of shear measured by the velocity gradient, G (units of s^{-1}) for given time periods, t (units of s). The unitless parameter Gt often is used to quantify the total shear applied during conditioning and was used in this study. As in previous research, shear was found to have a significant impact on polymer demand; as shear increased, polymer demand increased proportionally. Different digestion processes resulted in different shear sensitivities for the resultant sludges. Thermophilic processes, both anaerobic and aerobic, were the most shear sensitive, followed by temperature phased anaerobic digestion (TPAD) which utilizes thermophilic and mesophilic digestion in series. Mesophilic digestion was the least sensitive to shear of the different digestion types. Conventional aerobic and anaerobic digestion processes had similar shear sensitivities. Different polymer types also responded differently to shear. Low molecular weight polymers were not able to withstand significant amounts of shear, highlighting the importance of polymer bridging for good floc formation. The charge density and structure of the polymers also did not provide a clear-cut impact on OPD or dewatering properties for higher shear dewatering. This is contrary to current beliefs since structured polymers have been marketed as having greater shear resistance compared to non-structured polymers. One possible reason is that the shear imparted to the liquid does not fully represent the total shear experienced by sludges during conditioning and dewatering. For example, it was found in this study that cake shear had a different impact on the final cake characteristics compared to the liquid shear applied during conditioning. When shear was applied to the cake, the cake had an odor production profile much like a high solids centrifuge. However, when a large amount of shear was applied to the liquid during conditioning, the resulting cake did not produce odorous compounds. Currently no methods exist to quantify shear during dewatering, but additional research would be beneficial in this area. This research also quantified the equivalent shear of full-scale dewatering devices using the Gt approach. High solids centrifuges imparted the greatest amount of shear, followed by medium solids centrifuges and belt filter presses imparted the lowest shear during dewatering. One benefit of higher shear was that the cake solids increased with greater amounts of shear.

A mathematical empirical model was developed which took into consideration the biocolloid concentration, polymer charge density, and the shear. Based on numerous investigations and 66 different data sets, the simplest model with the highest correlation coefficient was a linear, multiparameter model of the following form:

$$\text{Polymer Dose (in meq/kg DS)} = -7.61 + 0.00692(\text{SCOD}) + 16.0(\text{PCD}) + 3.99\times10^{-4}(\text{Gt})$$

Where SCOD is the COD concentration in mg/L for filtrate less than 4.2 μm; PCD is the polymer charge density in meq/g; and Gt is the unitless mixing intensity parameter. The correlation coefficient, r^2, for this equation was 0.73, indicating a reasonably good fit. This model is the first that attempts to predict polymer demand given sludge, polymer and shear characteristics. The model was applied to a full-scale trial, and the predictions were very good for two different polymers and two different sludge types dewatered using a medium solids centrifuge. However, the model generally over-predicted polymer demand for the high solids centrifuges. Additional research could be used to refine the model, especially to better understand shear during dewatering and the interaction of polymer and solids types.

The model and the research also were used to demonstrate methods to reduce polymer demand, such as addition of inorganic coagulant and changes to centrifuge operation. For example, when the polymer was added directly into a full-scale centrifuge, the polymer demand was 18% less compared to adding the polymer approximately 20 m upstream of the centrifuge. The cake solids contents were approximately equivalent for the different locations. Reducing the torque of the centrifuge also reduced polymer demand, but reduced the cake solids content. Similarly, addition of ferric chloride prior to polymer addition reduced the polymer demand through a reduction in the biocolloid concentration.

CHAPTER 1.0

BACKGROUND AND OBJECTIVES

1.1 Introduction

Thickening and dewatering are used to reduce the volume of biological sludges. In general, the flocs that form in activated sludge systems and after both aerobic and anaerobic digestion will contain significant quantities of water. These flocs typically will not thicken and dewater well without conditioning chemical addition.

Conditioning and dewatering historically have been accomplished using ferric chloride, with or without lime, in dewatering equipment such as vacuum filters and recessed chamber filter presses. Subsequently, cationic polymers were introduced for sludge conditioning. These polymers are much easier and more convenient to use. The current technology for dewatering equipment, such as belt filter press and dewatering centrifuges, has performed well with cationic polymer conditioners.

Polymer conditioners are available in different formulations of shape, size and charge characteristics. For an excellent review of polymers, their characteristics, chemical makeup and usage, see Dentel (2001). The centrifuge and belt filter presses have become standard equipment for dewatering sludges and synthetic organic polymers are now overwhelmingly used as conditioners of activated and digested biosolids. Charge neutralization and polymer bridging are key mechanisms for conditioning, and the optimum polymer dose (OPD) typically occurs when the particle charge has been neutralized (Dentel et al., 2000; Abu-Orf and Dentel, 1997). Charge neutralization can result with the addition of cationic polymers that neutralize the negative charges in solution and on the floc. The negative charge in sludge originates from the organic molecules produced by microbes. For example, characterization of extracted biopolymer has shown that a portion of the polysaccharides are made up of uronic acids (Frolund et al., 1996; Forster and Lewin, 1972; Horan and Eccles, 1986; Brown and Lester, 1980), which have a carboxyl group that is unprotonated at typical pH ranges found in activated sludge systems and therefore contribute to the negative charge of bioflocs. In addition, the exocellular protein is rich in amino acids containing carboxyl groups such as glutamic and aspartic acid which will also contribute to the negative charge of the flocs (Higgins and Novak, 1997b; Dignac et al., 1998). Typically, the reduction of particle charge will decrease the repulsive forces between particles and promote aggregation.

Although it is generally accepted that charge neutralization is the main mechanism for conditioning by polymers, it is not well understood which sludge constituents create polymer demand. Also, a significant degree of variance has been observed in the polymer requirements for different types of digestion processes as well as for sludges generated using the same process but at different locations. The polymer demand for anaerobically digested biosolids in one treatment plant may be quite different from the anaerobically digested biosolids in another plant. To better understand this variability, there is a need to examine the interactions between the cationic polymer and sludges and the mechanisms of polymer conditioning. In addition, there are currently no methods available to predict polymer demand for different types of sludges, different types of polymer, and different dewatering devices. The polymer dose required for a particular sludge typically is estimated in the lab using bench-scale tests, such as capillary suction time (CST) and specific resistance to filtration (SRF), or in the field by visual observation of cake and filtrate/centrate properties.

In addition, polymers are expensive and their costs can represent a significant fraction of operation costs at treatment facilities. Therefore, improvements to bioflocculation and a better understanding of sludge properties that create polymer demand could result in decreased polymer demand for conditioning while achieving improved cake solids, throughputs and captures. These improvements will decrease operational costs.

A hypothesis is provided below to explain the cause of polymer demand in biological suspensions.

1.2 Hypothesis for Interaction Between Polymer and Bioflocs (Biocolloid Theory)

Sludges consist of organic, inorganic and microbial matter held together in a matrix known as a floc. These flocs are held together by exocellular biopolymers, mainly protein and polysaccharide, that are negatively charged. According to Frolund et al. (1996), up to 70% of activated sludge flocs can be biopolymer. These biopolymers typically have negatively charged functional groups which contribute to the overall negative charge of the particles (Dignac et al., 1998; Higgins and Novak, 1997a; Forster and Lewin, 1972). Though most of the biopolymer remains within the floc, pieces of the floc can become dislodged. These unattached particles are mainly biopolymer and will be referred to as biocolloids. The detachment of biocolloids from the floc can be represented in the form of a reaction as shown in Figure 1-1. This reaction can be considered a dynamic equilibrium, in that the amount of biocolloids can be influenced by wastewater composition, cation concentrations, mixing, SRT, and other factors.

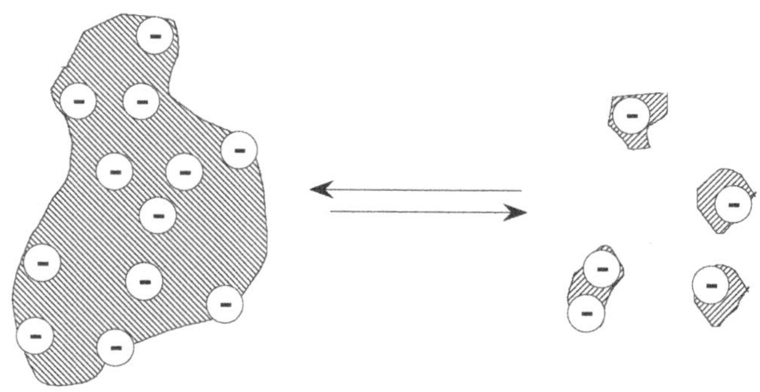

Figure 1-1. Reaction 1 Showing the Unattached Biocolloids and the Floc in Solution.

Charge neutralization is the main mechanism for conditioning, and the OPD typically occurs when the particle charge has been neutralized (Dentel et al., 2000). The amount of polymer required for effective dewatering is dependent on the quantity required to obtain a zero charge on the anionic colloidal particles. Thus, the polymer dosage depends on the solids characteristics and polymer because of the importance of polymer-particle interactions (Abu-Orf and Dentel, 1997).

As the equilibrium in Reaction 1 shifts towards the right, there is an increase in biocolloid concentration. It is likely that addition of cationic polymer to a sludge suspension coagulates these unattached anionic biocolloids and that much of the polymer demand is consumed by these biocolloids. The colloidal fraction has been shown to create the greatest difficulty in dewatering (Karr and Keinath, 1978) and therefore, removal of colloids by conditioning would enhance dewaterability. In addition, colloids have a very high surface area to mass ratio and assuming equal distribution of anionic sites they would create a significant amount of polymer demand. The interaction between cationic polymer and biocolloids is shown in Figure 1-2 and will be referred to as Reaction 2.

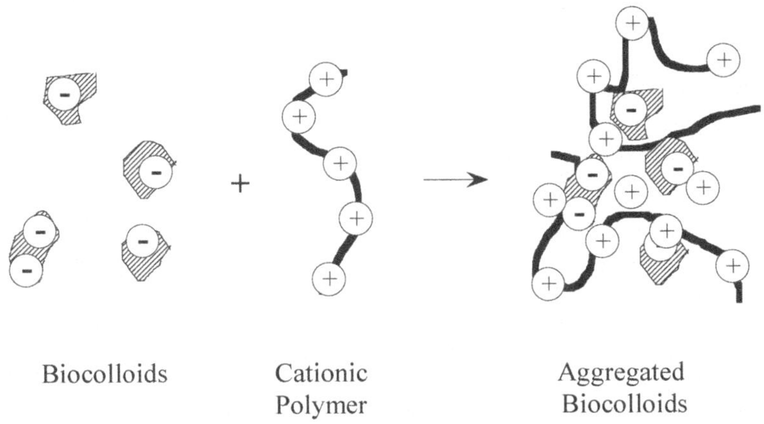

Figure 1-2. Reaction 2, Depiction of Unattached Biopolymer and Cationic Polymer Resulting in Coagulation. (Note: Not to Scale).

The reaction between the cationic polymer and anionic colloids creates a gel-like three-dimensional structure, enclosing the macroscopic particles, which dewaters readily (Roberts and Olsson, 1975). Reactions 1 and Reaction 2 form the basis of the proposed hypothesis that the biocolloids present in the solution are responsible for the polymer demand in different kinds of sludges and factors that affect the biocolloids concentration will affect the polymer demand. The amount of polymer required for effective conditioning depends on the biocolloid concentration. The higher the concentration of biocolloids, the greater the surface area and higher the polymer demand due to the increase in surface area and surface charge to be titrated.

The biocolloid theory for polymer demand comes out of work by Murthy and Novak and co-workers (1999 and 2000a), who reported a relationship between solution phase protein/polysaccharide and polymer demand. However, their studies were limited to a few sample types and this hypothesis has not been verified for a range of sludge types.

1.3 Summary of Research Needs

Based on data from the literature and from experiences at wastewater treatment plants, research is needed to answer the following questions:

1. Why do different sludge types have different polymer demand?
2. Why do different polymer types have different polymer demand?

3. Why do different dewatering devices have different polymer demand?

Previous research suggests that biocolloids present in the solution may be responsible for the majority of polymer demand. However, this has only been tested in limited numbers of samples. No research has been performed to examine the differences in polymer demand for different sludges from different processes and from different plants using the same digestion process. Thus, research is needed to examine these interactions and verify the hypothesis that biocolloid concentration is responsible for the majority of the polymer demand in different kinds of sludges—such as anaerobically digested biosolids, aerobically digested biosolids, autothermal thermophilic anaerobic digestion (ATAD) systems, returned or waste activated sludges and mixed liquor—that are produced under different conditions. Verifying the biocolloid theory will provide an important step in understanding the stoichiometry of polymer conditioning and the factors which affect polymer demand.

In addition, a method or model to predict polymer demand based on fundamental characteristics (e.g., biocolloid concentration, polymer characteristics and dewatering device) would be useful to utilities and practitioners. Methods to reduce the biocolloid concentration should the optimum polymer dose and lower costs associated with polymer addition.

1.4 Research Objectives

The overall goal of this research is to establish the stoichiometry for cationic polymer conditioning, provide a practical understanding of factors that affect polymer demand, and allow prediction of polymer demand for a given sludge type, polymer, and dewatering device.

The primary objectives can be broken down into the following:

- Verify the hypothesis that biocolloids present in suspension are responsible for the majority of polymer demand.
- Understand the differences in polymer demand for different polymers.
- Understand the impact of shear on polymer demand.
- Combine the factors of biocolloid concentration, polymer characteristics, and shear into a model that can be used to predict polymer demand for different treatment scenarios.
- Apply the model to predict polymer demand for different treatment, conditioning and dewatering scenarios.
- Apply the knowledge gained to develop methods to reduce polymer demand.

CHAPTER 2.0

METHODS AND MATERIALS

2.1 Experimental Approach

To study the impacts of three parameters—sludge characteristics, polymer characteristics, and shear —on polymer demand, experiments were conducted in which only one variable or parameter was varied at a time. For example, to study the impact of sludge characteristics on polymer demand, a number of different kinds of sludges (such as anaerobically digested sludge, aerobically digested sludge and autothermal thermophilic aerobic digestion (ATAD) sludge) were obtained from different treatment plants. The samples were dewatered using the same mixing protocol and the same polymer type and concentration. In addition, the sludges were characterized for a number of different parameters which typically included:

1. total suspended solids (TSS);
2. volatile suspended solids (VSS);
3. soluble chemical oxygen demand (SCOD);
4. soluble protein concentration;
5. soluble polysaccharide concentration;
6. capillary suction time (CST);
7. particle size distribution;
8. soluble and total metal concentrations;
9. ammonium concentrations;
10. amino acid analysis;

Using this information, relationships between sludge properties and the polymer demand were developed. Similar approaches were used to study polymer characteristics and shear.

2.2 Analytical Approach and Methods

The analytical methods used to determine the above-mentioned parameters are explained below.

TSS and VSS. TSS and VSS were measured using Method 2540D and Method 2540E of *Standard Methods* (APHA et al., 1995).

Filtration. Samples were centrifuged at 3000 x g for 15 minutes and the centrate was filtered using a Millipore AP40, 47mm filter with a pore size of 2.4 to 6.0 μm. This filtrate was defined as the "soluble fraction" and was used for the following tests.

Soluble chemical oxygen demand. The soluble fraction was analyzed for SCOD using Hach COD vials.

Soluble protein concentration. The soluble fraction was analyzed for proteins using the Hartree (1972) modification of the Lowry et al. (1951) method. Protein standards were prepared using bovine serum albumin.

Soluble polysaccharide concentration. The soluble fraction was analyzed for polysaccharide using the method of Dubois et al. (1956) using glucose as a standard.

Capillary suction time (CST). Samples were analyzed for CST using a Triton-Type 165 CST apparatus and Whatman # 17 Chromatography paper according to Method 2710G of *Standard Methods* (APHA et al., 1995). One CST paper was used for each test. Lower CST values are associated with better dewatering properties.

Polymer solution. A known concentration of cationic polymer was prepared with a target strength of 0.5% by weight. The neat polymer was weighed and added to 1000 mL of tap water and the solution was mixed for 30 minutes at 500 rpm. Using the weights of the polymer and water, the concentration of the polymer solution was determined.

Optimum polymer dose. In these tests, a known mass of polymer was mixed with a known amount of sample using defined mixing protocols. After mixing, the CST was measured as a method to quantify dewatering properties. The optimum polymer dose (OPD) was defined as the polymer dose that produced a minimum CST (Novak and Lynch, 1990; Dentel et al., 2000; Novak et al., 1988).

Particle size analysis. Particle size analysis was carried out using Met One Water grab sampler capable of counting the particles ranging from 2 to 60 μm.

Cake solids. Samples were obtained by removing a portion of the sludge cake formed on the belt filter press. Sludge samples were placed on the pre-weighed tin and immediately weighed to reduce the moisture loss. Samples were dried at 105 ^{o}C for 2 hours, desiccated and weighed. Cake solids concentrations were expressed in percentages.

Total and soluble metals analysis. The total metals content (including Fe, Al, Ca, Mg, Na, and K) was determined from the samples after concentrated nitric acid-heat digestion followed by analysis using inductively coupled plasma (ICP), Method 3500. The digestion procedure used was Method 3030 E in *Standard Methods* (1999). The soluble fraction was measured using the filtrate (described above) and analyzed for Ca, Mg, Na, and K by ICP according to Method 3500.

Ammonium analysis. The soluble samples were filtered a second time through a 0.45 μm filter and analyzed for ammonium using a Dionex Ion Chromatograph with a CS12 column and conductivity detection, with self-generating suppression of the eluent. The eluent was 20 mM methanesulfonic acid, introduced at a flow rate of 1mL/min.

Amino acid analysis. Individual amino acids were measured using High Performance Liquid Chromatography (HPLC) subsequent to derivitization.

Shear experiments. A series of experiments were performed to investigate the effect of shear on polymer demand and solids characteristics. The shear or mixing intensity is typically quantified using the velocity gradient, G (s^{-1}) multiplied by the time of mixing. This unitless parameter, or Gt value, can be used to examine the impact of shear on solids properties and polymer conditioning (Novak and Lynch, 1990; Novak et al., 1988; Werle et al., 1984). The G value for a given mixer setup (paddle, reactor vessel, and speed) can be determined by creating a curve of G versus mixer speed for the mixing regime. Typically, torque is measured and from this, G is calculated (see Werle et al., 1984). The velocity gradient as a function of mixer rpm was determined for the mixer setup at Bucknell University using a torque readout from the mixer. The mixer was a Labmaster Lightning mixer. The net torque values were determined by subtracting the torque during mixing in air from the torque measured during mixing of the solids sample. The torque measured by the mixer in air was considered due to internal friction of the mixer setup. The mixer setup used a 2-L, baffled mixing vessel with 500 mL of sample and a flat paddle blade. As an example of a typical scenario, to achieve a G of 1000 s^{-1} using a 1000 mL sample, the mixer speed required was 564 rpm. Therefore, to attain a Gt value of 100,000, the sample would be mixed at 564 rpm for 100 s.

Belt filter press simulation. Galla et al. (1999) have shown that the belt filter press simulator (BFPS) or crown press simulator correlated well with the full-scale belt filter press results. A Phipps and Bird crown press simulator was used to simulate the belt filter presses used in treatment plants. The simulator was used to provide the data on relative cake solids moisture content and filtrate clarity for different sludges. A 150 mL solids sample was applied on the belt, and allowed to free drain for 60 seconds. Next, the solids were formed into a cake of about 10 cm diameter with uniform height, and pressure of 200 psi was applied for 60 seconds. The cake was removed and analyzed for solids and moisture content.

Vacuum filtration for cake solids. A second solids sample was dewatered using vacuum filtration and a Buchner funnel with Whatman#2 filters. A 50 mL sample was placed in the Buchner funnel, and a constant pressure of 15 inches of Hg was applied for 60 seconds. After this time period, the vacuum was shut off, and the cake was removed and analyzed for solids and moisture content.

Polymer charge density. Polymer charge density was measured using the colorimetric titration approach described by Dentel et al. (1993) and Kam and Gregory (1999).

CHAPTER 3.0

SLUDGE PROPERTIES IMPACTING POLYMER DEMAND

3.1 Objective I—Verifying the Biocolloid Theory—Polymer I

3.1.1 Overview of Study

One of the main objectives of this research was to verify the hypothesis that biocolloids in solution are responsible for the majority of the polymer demand during conditioning, and in effect try to determine the stoichiometry for polymer demand. A total of 19 samples were tested for this initial task. These samples include mixed liquor, RAS, aerobically and anaerobically digested, a combination of aerobic/anaerobic digestion, a thermophilic anaerobic digested, and an ATAD sample. The OPD, SCOD, protein and polysaccharide concentrations are shown in Table 3-1. The OPD was determined using the same polymer and same testing conditions each time (NALCO Cationic Polymer 7194). The data in Table 3-1 have been sorted from lowest to highest polymer demand to facilitate comparison of the parameters in the table.

Samples containing the lower polymer demand generally were aerobic samples, such as RAS, MLSS or aerobically digested. The anaerobically digested samples (which made up the bulk of sampling) had a fairly large range in polymer demand from 5.6-12.7 g/kg and generally had greater polymer demand than the aerobically digested samples. In other words, biosolids using the same digestion process had polymer demand that could vary by a factor of two. The polymer demand was determined using the same procedure and polymer for each sample, so the differences can be attributed to sludge properties. The two thermophilic samples had the greatest polymer demand of all the samples tested, about 4-5 times that of the mesophilic digested products.

The relationships between the different parameters measured and how these relationships fit with the biocolloid theory are examined in the following sections.

Table 3-1. Summary of Results from Batch Sampling Tests.

Sample Number and Description	OPD (g/kg DS)	SCOD (mg/L)	Average Soluble Protein (mg/L)	Average Soluble Polysac. (mg/L)	Soluble Prot+Polys (mg/L)
1-MLSS	3.9	29	1	7	7
2-Aerobic/Anaerobic Digestion	4.2	706	135	71	206
3-Anaerobic Digestion	5.6	780	387	76	463
4-RAS	6.1	51	10	8	8
5-Anaerobic Digestion	6.2	762	570	62	287
6-Anaerobic Digestion	6.6	837	247	64	311
7-Aerobic Digestion	7.2	852	132	119	251
8-Anaerobic Digestion	8.4	1310	609	76	685
9-Anaerobic Digestion	8.8	1365	629	115	744
10-Anaerobic Digestion	9.1	1048	285	51	336
11-Anaerobic Digestion	10.1	2596	1038	151	1189
12-Anaerobic Digestion	10.2	1898	849	177	1026
13-Anaerobic Digestion	10.5	1173	523	91	613
14-Anaerobic Digestion	10.8	2546	1155	178	1333
15-Anaerobic Digestion	11.0	1860	1063	129	1192
16-Anaerobic Digestion	11.3	2522	1297	144	1441
17-Anaerobic Digestion	12.7	2540	1295	180	1475
18-Anaerobic Digestion-Thermophilic	37.6	2804	1774	219	1993
19-ATAD	56.4	3876	1507	236	1743

3.1.2 Role of Protein

The central tenet of the biocolloid theory is that small particles, made up mainly of biopolymer, create a significant amount of polymer demand. To further examine this hypothesis, the OPD data are correlated to the biopolymer fractions in the following sections. As seen in Figure 3-1, most of the samples showed a linear correlation between the protein concentration and OPD, except the two thermophilic samples (18 and 19). This trend occurred for many parameters in that a relationship was found between the OPD and other parameters and the two thermophilic samples were outliers. The correlation of OPD and protein concentration for all the samples except the thermophilic samples is shown in Figure 3-2.

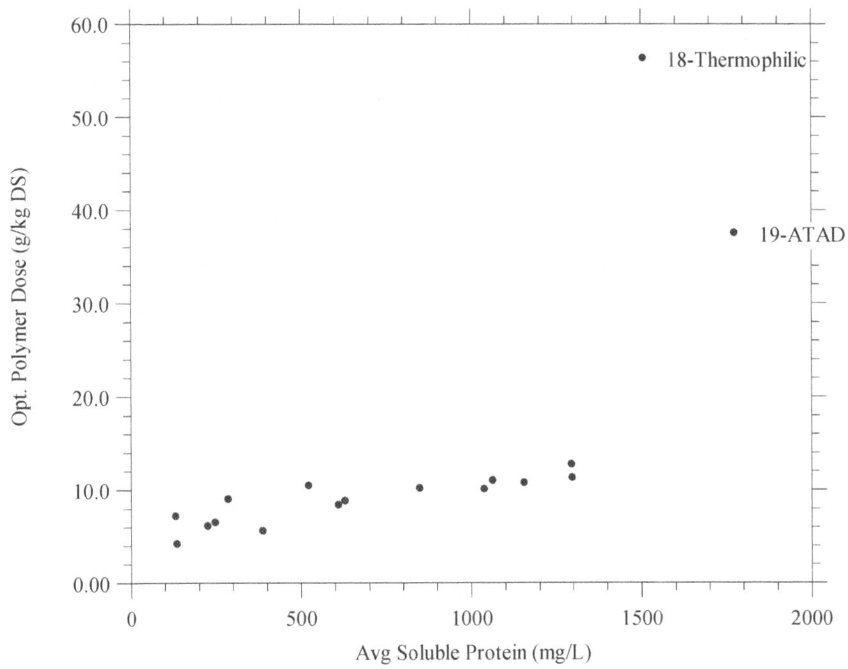

Figure 3-1. Relationship Between Optimum Polymer Dose and Soluble Protein Concentration for all Samples.

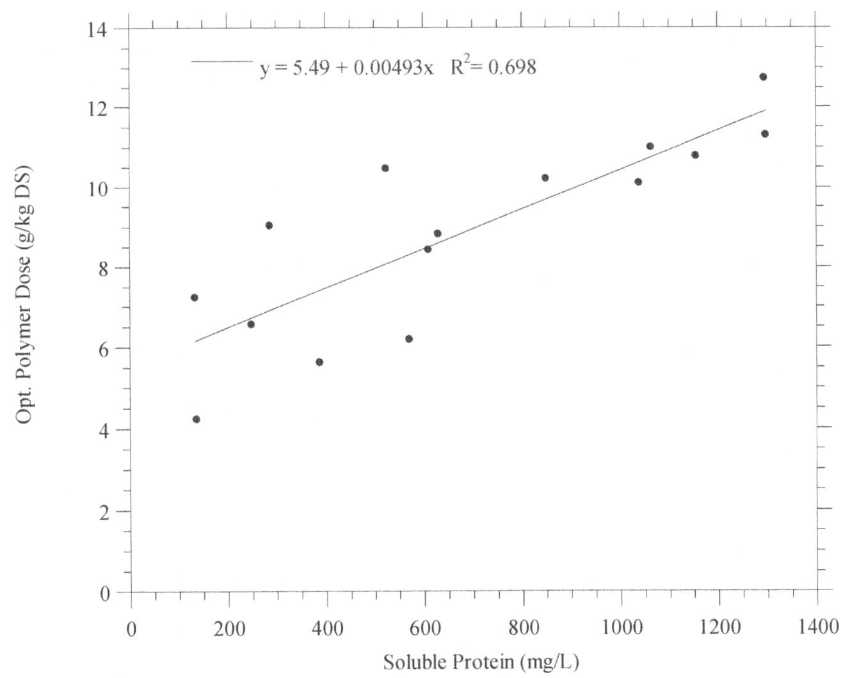

Figure 3-2. Relationship Between Optimum Polymer Dose and Soluble Protein Concentration for all Samples Except Thermophilic Samples.

The linear correlation between this parameters is reasonably good, with an r^2=0.70. The relationship for the anaerobically digested samples only (Figure 3-3) shows a reasonably good correlation between the OPD and the soluble protein concentration (r^2=0.64).

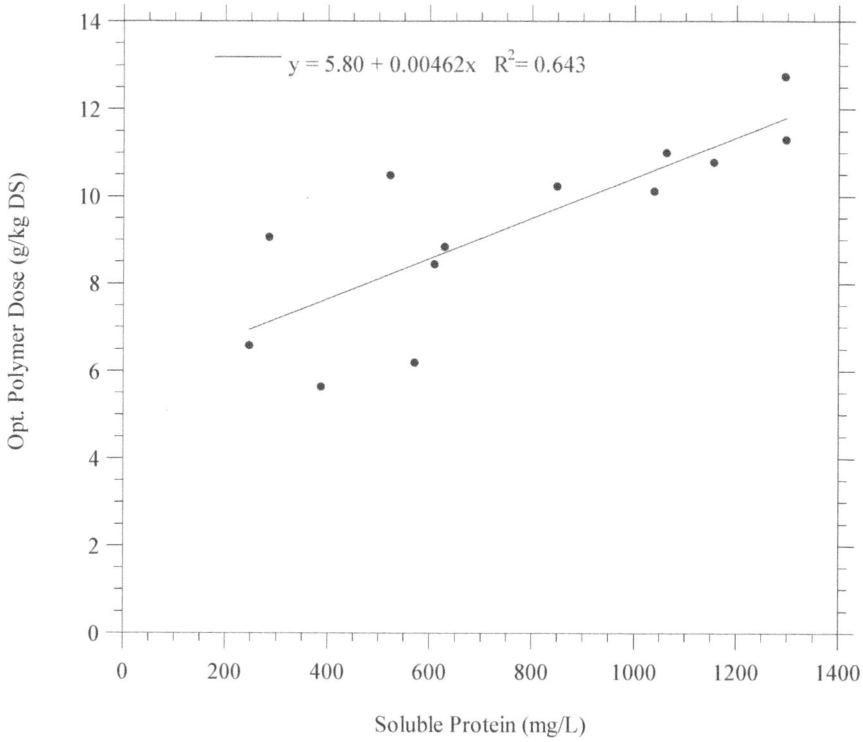

Figure 3-3. Relationship Between Optimum Polymer Dose and Soluble Protein Concentration for Anaerobically Digested Samples Only.

The results from this analysis help to confirm that biocolloids create polymer demand. These biocolloids, made up of mainly protein and polysaccharides, would be expected to contain significant amounts of negatively charged functional groups. For example, the amino acids, aspartic acid (ASX) and glutamic acid (GLX) have carboxyl groups that would be unprotonated at pH ranges typical of these systems. Similarly, polysaccharides generally have uronic acids which also contain carboxyl groups that would contribute to the negative charge of the particles. An intensive study of 10 of the anaerobically digested samples was undertaken and the amino acid composition was analyzed. The positive correlation between OPD and ASX or GLX (Figure 3-4) supports the concept that the negatively charged groups on the protein create the polymer demand.

It would also be interesting to confirm if the colloidal protein has different amounts of these carboxylated amino acids which would help explain, in part, the differences in polymer demand for different samples. A plot of soluble protein versus GLX and ASX (Figure 3-5) shows a strong linear correlation between ASX/GLX and the soluble protein fraction. This suggests that the soluble protein characteristics are similar for the different samples, and the relative amounts of the carboxylated amino acids from the different samples are equivalent. Therefore, the differences in polymer demand are likely more a function of solution biocolloids quantity since the characteristics of the biocolloids appears to be similar.

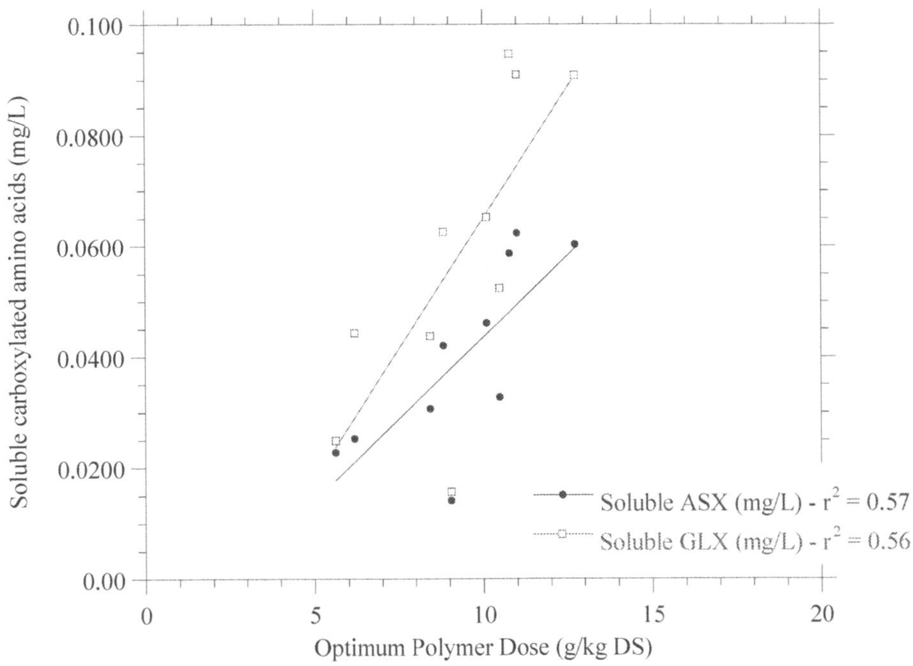

Figure 3-4. Relationship Between ASX and GLX and OPD for Anaerobically Digested Samples.

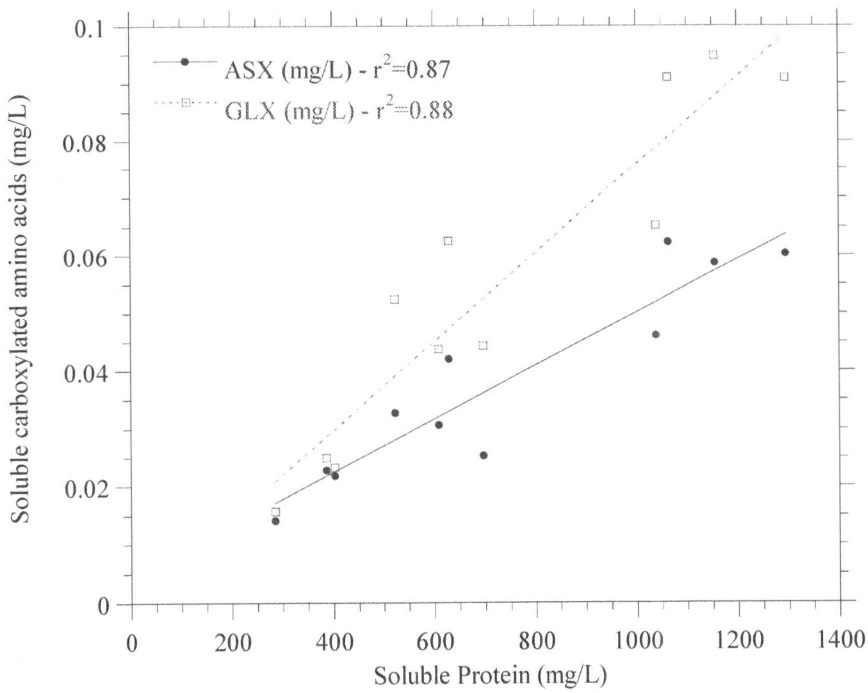

Figure 3-5. Relationship Between GLX and ASX and Soluble Protein.

3.1.3 Role of Polysaccharides

Since polysaccharides also can contribute negatively charged groups to the biocolloids, it is likely that the polysaccharides also contribute to the polymer demand. Therefore, considering both protein and polysaccharides together may produce better correlations. Polysaccharide concentrations for the samples are provided in Table 3-1. A plot of OPD vs. the sum of polysaccharide and protein concentration is shown in Figure 3-6. This graph is for all samples except the two thermophilic samples. The results show a stronger correlation (r^2=0.76) for OPD compared to protein alone. For the anaerobically digested samples only, the r^2 was 0.67, which is a slight improvement over protein alone, Figure 3-7.

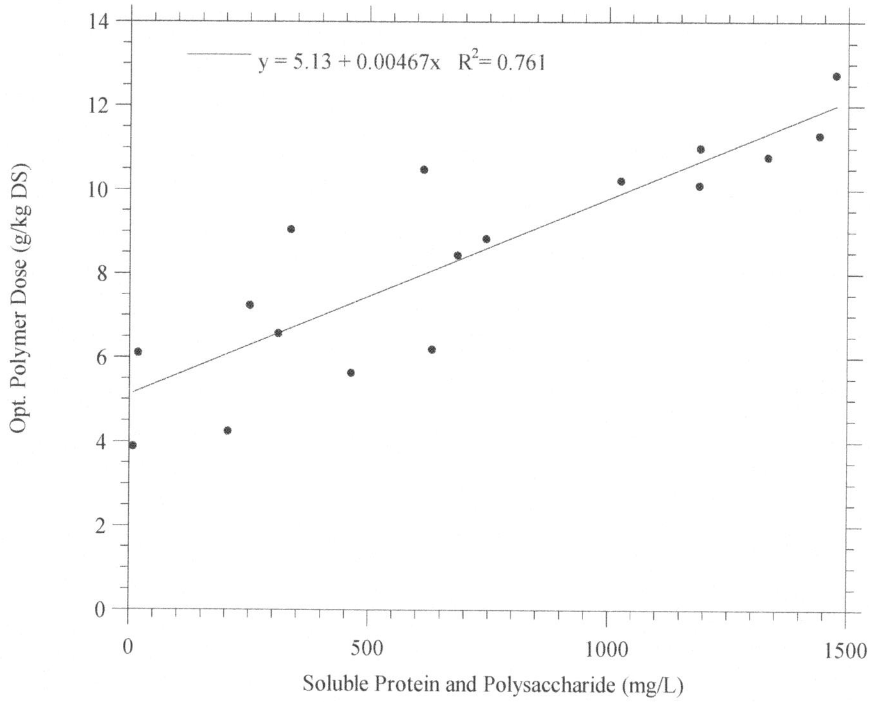

Figure 3-6. Relationship Between OPD and Solution Protein Plus Polysaccharide Concentration for all Samples Except Thermophilic.

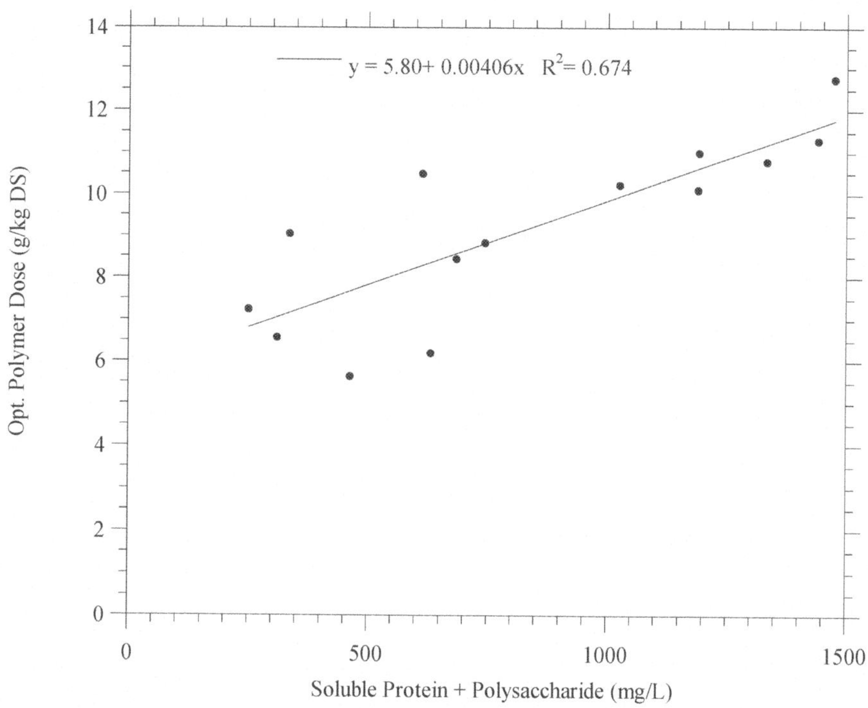

Figure 3-7. Relationship Between OPD and Solution Protein Plus Polysaccharide Concentration for Anaerobically Digested Samples.

Therefore, the two main components of biopolymer, protein and polysaccharide, can explain much of the differences in polymer demand in sludge samples. No significant correlations were found between the biopolymer fractions bound to the floc particles (extracted with buffer and shear). Rather, it is the smaller biopolymer fraction (<6.0 µm) that is related to the polymer demand and not the larger particles. As additional confirmation that biocolloids create polymer demand, the concentration of solution protein and polysaccharides was measured during conditioning for a number of samples. A typical data set is shown in Figure 3-8. As more polymer was added, the concentration of the biocolloids, as measured by protein and polysaccharide, decreased. The CST followed the classic dose-response curve in that it decreased, reached a minimum, and then began to increase as the suspension became overdosed. The solution protein and polysaccharide concentration continued to decrease, but slightly, even as the CST began to increase, suggesting that the polymer is reacting with the biocolloids and sequestering the protein and polysaccharide into the larger flocs. The OPC seems to occur soon after these biocolloids have been largely removed from suspension. For example, in Figure 3-8, the optimum dose occurs after the inflection point in the protein and polysaccharide concentration curves. The protein and polysaccharide remaining in solution could be small sized material that coagulates poorly, so additional polymer dosing does not effectively remove this material.

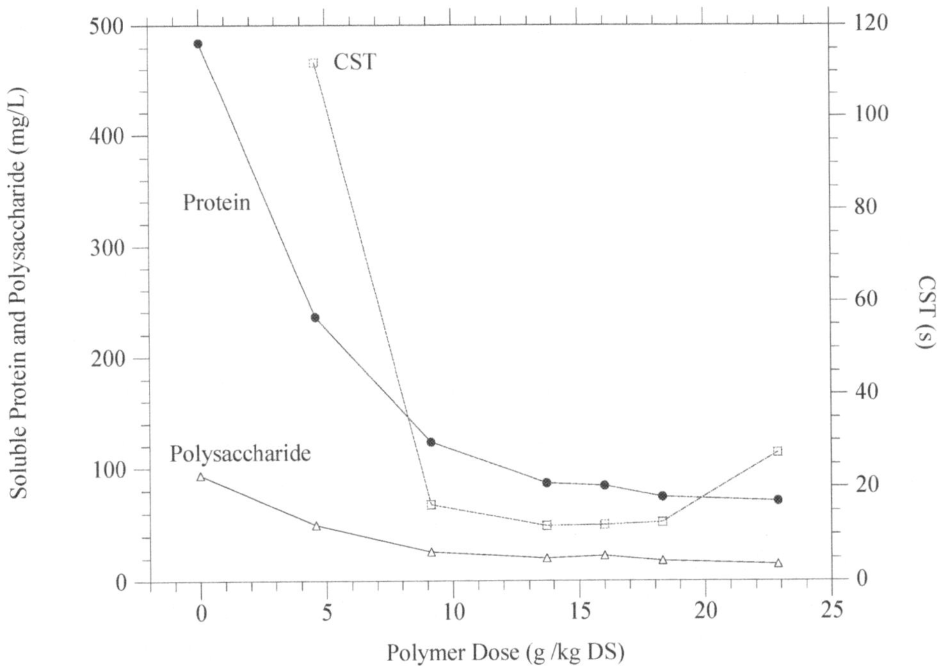

Figure 3-8. Soluble Protein and CST as a Function of Polymer Dose.

The data presented in the previous paragraphs support the hypothesis that biocolloids create polymer demand, and greater amounts of biocolloids are directly related to greater polymer requirements. The implication of these results for process operations is that minimizing the quantity of these small particles should reduce dewatering and disposal costs.

It is also interesting to note that the y-intercept for these graphs is between 4-6 g/kg DS. This suggests that the samples have an inherent demand related to the non-biocolloid fraction, such as the larger particles that are removed by the filtration step. It is likely that the larger particles have the same charge characteristics as the biocolloids since the biocolloids are in fact smaller pieces that have broken off the larger flocs. For example, Figure 3-5 shows two of the charged groups within protein were relatively constant as a fraction of the protein concentration. In other words, the protein had the same percentage of charged groups, and the variability between samples was due to the total amount of the protein, not the characteristics of the protein. This inherent demand due to the larger particles is about 5 g/kg DS (the y-intercept value).

It is not clear why the two thermophilic samples were outliers in the correlations between biocolloids and OPD. Based on the data for the other sample types, the thermophilic samples have much less biocolloid content than would be expected for its polymer demand, or its polymer demand is much greater than expected for its biocolloid concentration. There are several possible explanations. It is possible that the particle size distribution for the biocolloids in the thermophilic samples is smaller such that they have more surface sites per unit weight of protein and polysaccharides. This would result in greater polymer demand per amount of biocolloid than other samples. Smaller particle size distributions may be a result of higher temperatures which could aid in deflocculation. Continuing work on biocolloid particle sizes will aid in understanding this effect. It is also possible that some other constituent that is not measured exhibits polymer demand in these samples. In addition, the particles > 6.0 μm could have contributed significant polymer demand unlike the other samples. These results and size fraction analysis are examined more fully in Section 3.3. It is also possible that the results from the thermophilic samples are due to experimental error.

3.1.4 SCOD as a Surrogate Measure for Biocolloids

An important aspect of this research is to provide practical tools that the wastewater treatment industry can implement to improve and optimize operations. Since protein and polysaccharide analysis require a setup and reagents not typically used at treatment plants (although their analysis could be implemented without significant costs at most facilities), a simpler and cheaper measure would be beneficial. SCOD is a possible surrogate measure of protein and polysaccharide concentration. SCOD has the benefit of being commonly used at many treatment facilities and would therefore be easily implemented. Plots of SCOD versus the soluble protein and polysaccharide (Figure 3-9) and the sum of protein plus polysaccharide (Figure 3-10) show a very good linear correlation between protein and polysaccharide and the soluble SCOD concentration. In addition, the linear correlation between SCOD and OPD results in an r^2 of 0.75 which is slightly better than the protein and polysaccharide correlation for anaerobically digested samples. As a result, SCOD would be a good surrogate measure of these parameters for treatment facilities. This concept may seem inappropriate since most practitioners would consider the SCOD from a digester as a measure of the volatile fatty acids (VFAs) concentration. VFAs were also measured in the 10 anaerobically digested samples, and no significant correlation was found between VFAs and SCOD (r^2=0.20, data not shown).

Therefore, plants investigating methods to reduce the biocolloid concentration (for example, effect of detention time or chemical additives) could use SCOD as a simple measure of biocolloid concentration. Additional research should investigate other potential measures of biocolloids that could incorporate on-line systems for in-plant use such as total organic carbon (TOC) analyzers as well as UV 254/280 absorbance.

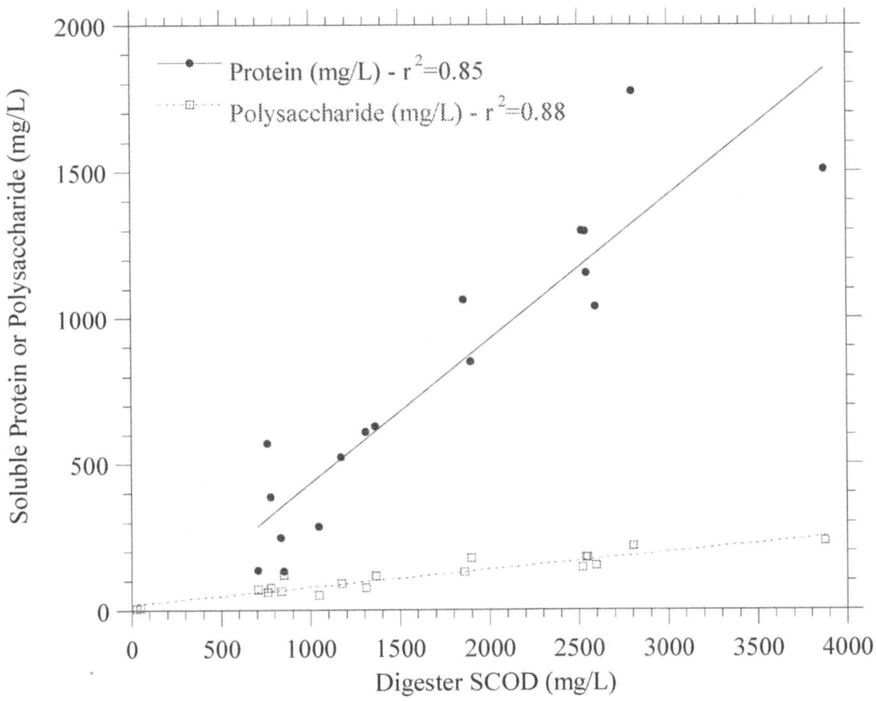

Figure 3-9. Relationship Between Solution SCOD and Soluble Protein and Polysaccharides for All Samples.

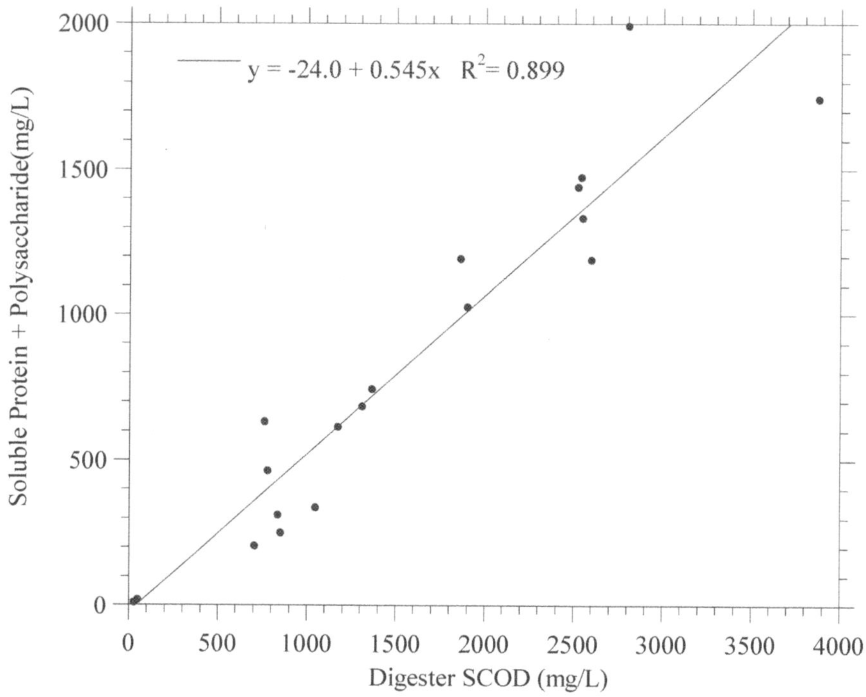

Figure 3-10. Relationship Between Solution COD and Soluble Protein and Polysaccharides for Anaerobically Digested Samples.

3.2 Impact of Other Sludge Characteristics and Process Parameters

Analysis of the data has shown that biocolloids, defined as particles smaller than 6.0 μm and made up of mainly protein and polysaccharide, appear to create the majority of the polymer demand. The OPD and biocolloid concentrations varied considerably for the different sample types, as well as for the same sample type (all anaerobically digested) from different locations. It would be important then to understand the cause of these differences in the biocolloid concentration to eventually lead to methods to reduce OPD by reducing the biocolloid concentration.

3.2.1 Role of Metals

Metals have been shown to play a key role in floc formation and the release/retention of biocolloidal materials (Higgins and Novak, 1997a, b). Both soluble and total metals were measured in the 10 mesophilic anaerobically digested samples and the one thermophilic anaerobically digested sample, and these metals included Fe, Al, Ca, Mg, Na, K. Another cation of importance is ammonium (NH_4^+) which was also analyzed in the soluble fraction.

The relationship between the total Fe concentration and the OPD is shown in Figure 3-11 for all these samples. Once again the thermophilic sample (#18) is an outlier for this evaluation. Figure 3-12 shows the same relationship for all the mesophilic anaerobically digested samples. These two parameters exhibited a strong negative correlation (r^2=0.86). As the total Fe concentration increased in the samples, the OPD decreased. There are several possible explanations for this result. First, since Fe will form mainly positively charged ions, Fe could reduce the charge of the particles and thereby reduce the polymer demand. Second, researchers at Aalborg University and Virginia Tech have shown Fe to be an important parameter in floc aggregation and retention of biopolymer (Nielsen and Keiding, 1998; Murthy et al., 2000b); therefore, the presence of greater amounts of Fe should reduce the solution phase biocolloids. To examine this second hypothesis, the soluble biopolymer concentration is plotted as a function of the total Fe concentration in Figure 3-13. Both protein and polysaccharide are negatively correlated with total Fe, and the correlation is stronger for protein than for polysaccharide concentrations. Therefore, it appears that Fe plays a role in reducing biocolloid concentration and thereby reducing polymer demand, and it is possible that Fe plays several roles in determining the OPD of sludge since the relationship between OPD and Fe is better than for Fe and biocolloid concentration. Also, other factors may impact the biocolloid concentration in addition to the Fe concentration.

These results are contradictory to a study by Feitz et al. (2002) who reported that greater amounts of iron in anaerobically digested sludge resulted in a deterioration in dewatering properties (measured by CST) due to more fine particles (<10 μm). However, the research in this study did not find a correlation between a sludge's initial CST and OPD, so it is difficult to determine from the Feitz et al. (2002) study whether Fe would have negatively impacted the OPD.

Although the literature is conflicting, the present results suggest that addition of iron to anaerobic digesters or just prior to polymer addition could improve dewatering by reducing polymer demand. Research by Murthy et al. (2000b) and Abu-Orf et al. (2001) showed that iron addition (in the form of ferric chloride, $FeCl_3$) reduced polymer demand of ATAD sludges through binding of biocolloids, mainly protein. Additional testing in this project examined the effect of iron addition on polymer demand and biocolloid sequestration, and the results are presented in Chapter 8.0. The results indicated that ferric chloride addition after digestion and before polymer addition was effective in reducing the biocolloid concentration, and thereby reducing the polymer demand. The addition after digestion may provide more predictable results compared to addition during digestion because other reactions occurring during digestion can influence iron chemistry. Reduction to ferrous iron and precipitation with sulfide species, for example, may reduce the ability of iron to bind biocolloids and reduce its impact on polymer demand. Other factors that may impact biocolloid concentration include SRT, mixing, other chemical additives, and small variations in temperatures.

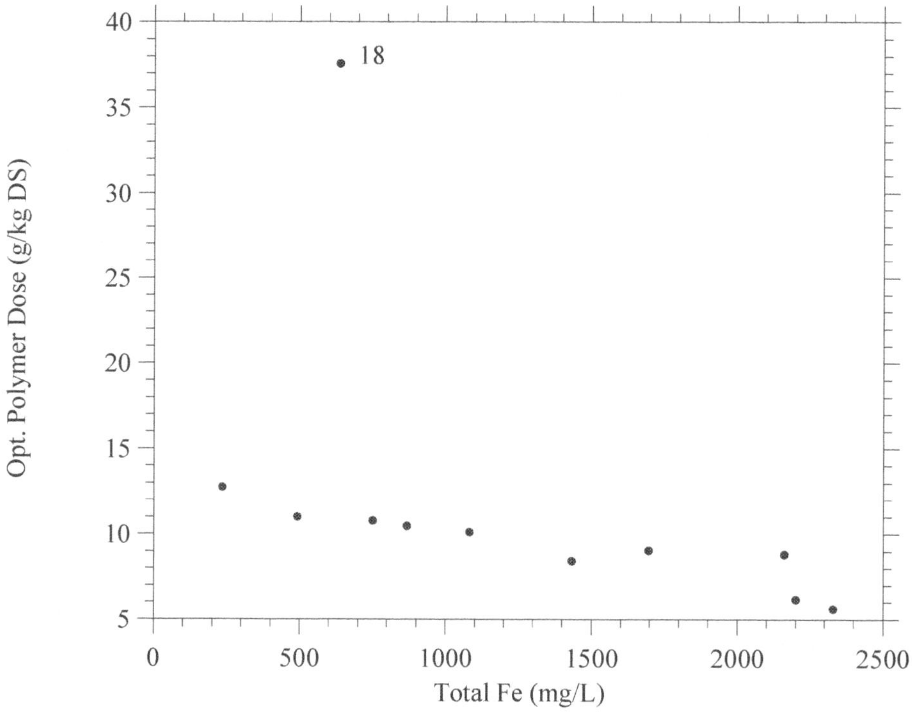

Figure 3-11. Relationship Between OPD and the Total Fe Concentration of Mesophilic and Thermophilic Anaerobically Digested Samples.

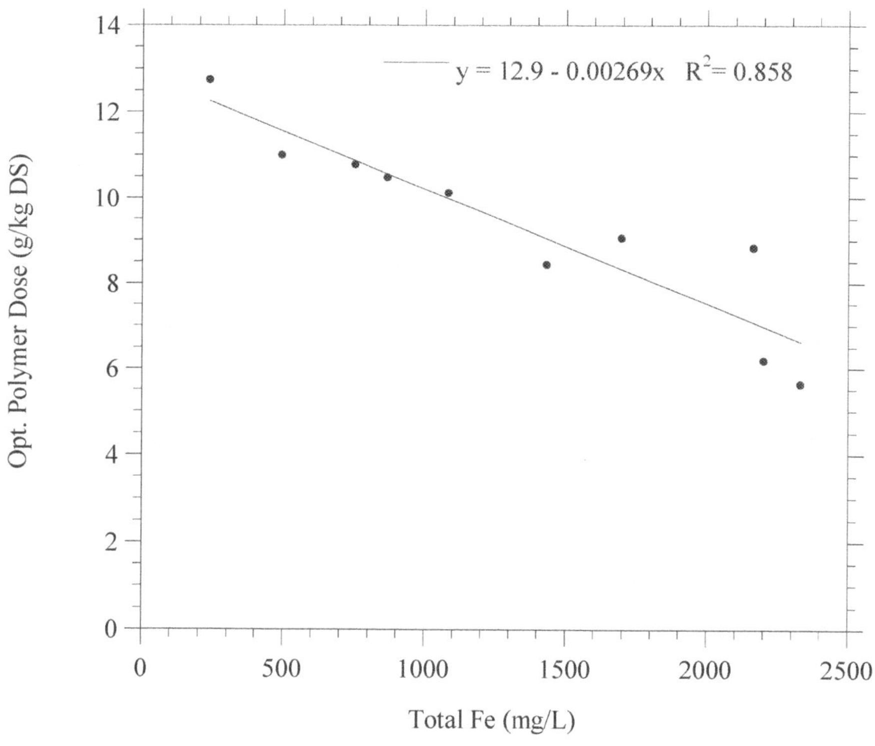

Figure 3-12. Relationship Between OPD and the Total Fe Concentration of Mesophilic Anaerobically Digested Samples.

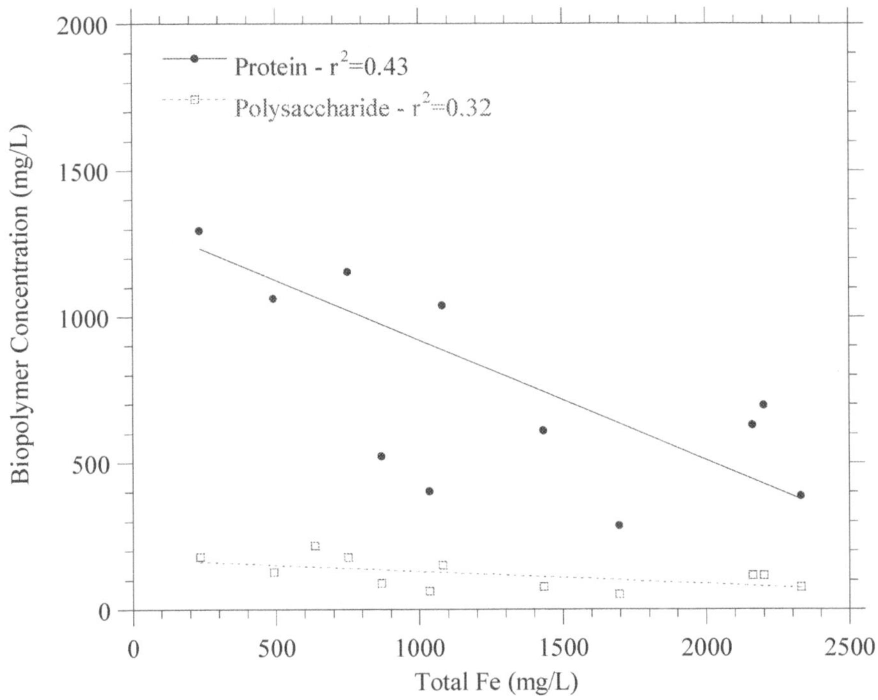

Figure 3-13. Relationship Between Soluble Protein and Polysaccharide Concentration and Total Fe Concentration in Anaerobically Digested Samples.

No other metals showed a significant correlation with polymer demand, including the monovalent-to-divalent ratio (M/D) proposed by Higgins and Novak (1997a). A summary of the linear correlation coefficients between the other cations and OPD is shown in Table 3-2. It is also important to note that not all correlations are necessarily linear. For example the effect of the M/D ratio has been shown to produce an optimum dose type profile, therefore a linear correlation of this parameter with OPD would not be appropriate. However, graphs of these parameters showed no apparent trends that would suggest a good non-linear relationship existed.

A plot of M/D ratio versus soluble protein plus polysaccharide for the anaerobically digested biosolids shoed a consistent relationship except for one outlier (Sample 17). When this point is omitted, the linear correlation coefficient is 0.60 showing some relationship between M/D and biocolloids for most anaerobic plants, although this did not translate into a significant correlation between M/D and OPD (r^2=0.20).

Table 3-2. Summary of Linear Correlations Between OPD and Cations.

Parameter linearly correlated against OPD	Correlation coefficient (r^2)
M/D ratio of soluble metals	0.013
Individual soluble cations (Na, NH_4, K, Mg, Ca)	all < 0.05
Individual total cations (Na, NH_4, K, Mg, Ca)	all < 0.12
Total Fe	0.86
Total Al	0.034

3.2.2 Effect of Process Parameters

Process parameters were investigated for the anaerobically digested samples to determine whether they affect polymer demand or biocolloid concentration. The parameters examined included SRT, volatile solids (VS) destruction, fraction of WAS in the digester feed, and time-temperature combinations. For these samples, no significant linear correlations were found (Table 3-3).

Table 3-3. Correlations Between Different Process Parameters.

Parameters	OPD (g/kg DS)	Prior Month Digester SRT (d)	Digester VS Destruction (%VS)	Loading Rate (kg VS/m^3-d)
OPD (g/kg DS)	1.000	0.030	0.147	0.290
Digester SRT (d)	0.030	1.000	-0.198	-0.525
Digester VS Destruction (%VS)	0.147	-0.198	1.000	-0.170
Loading Rate (kg VS/m^3-d)	0.290	-0.525	-.170	1.000

The lack of correlation could be due to the many different variables that affect polymer demand, such as Fe content, waste characteristics and biocolloid concentration at the different plants used in the study. A controlled laboratory study by Houghton et al. (2000) also showed that anaerobic digestion time did not have an impact on exocellular biopolymer concentration such as protein and polysaccharide.

3.3 Objective I—Verifying the Biocolloid Theory—Effect of Biocolloid Size Fractions

The previous results support the theory that biocolloids present in solution create polymer demand and a good correlation was seen between biocolloid concentration and optimum polymer demand, *except for* thermophilic samples. However, the theory does not seem to be universal since thermophilic samples did not fit this trend. As a result, the theory warrants additional analysis to determine if the relationship holds for other polymer types and if it can be adjusted to include high polymer demand samples such as the thermophilic samples. Results from particle size analyses showed that the thermophilic samples had high particle numbers in the 4 - 16 μm range compared to the other samples (Figure 3-14). The filters used in the previous sample analyses had a nominal particle size retention between 2.4 - 6.0 μm. Assuming a mean retention size of 4.2 μm, this filter may exclude particles that are important in determining the polymer demand. To examine the effect of a different polymer and test the hypothesis that particles in the 4 - 16 μm size range may better explain the polymer demand, additional tests were performed. These tests were performed using a linear polymer (Betz CE 2680) and the sample filtrate was collected after filtering through 4.2 mm (filter from previous experiments), 8 μm, 16 μm, and 22.5 μm filters. Filtration was performed on parallel filtration apparatus. In addition, the mixing regime was changed to a Gt of 20,000 using a standardized mixing apparatus (see section on impact of shear for mixer standardization details).

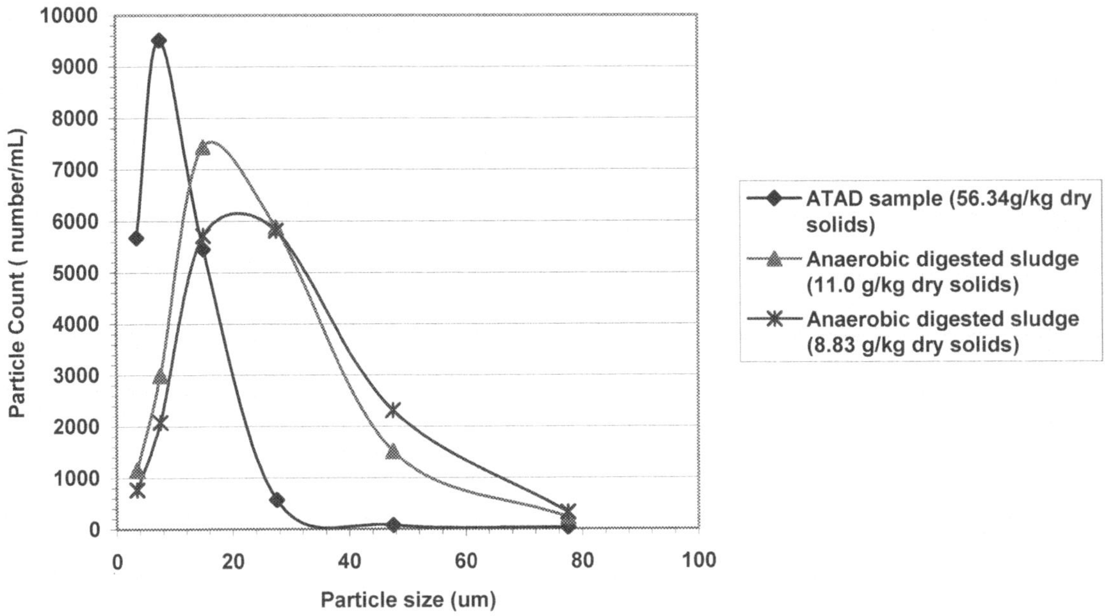

Figure 3-14. Particle Size Distribution of ATAD and Anaerobically Digested Samples.

A summary of the data from these trials is provided in Table 3-4. As shown in the table, 10 samples were tested, including one high polymer demand sample—the temperature phased anaerobic digestion (TPAD) sample. The correlation between the protein plus polysaccharide concentration and the OPD (Figure 3-15) is very good, and all size fractions provide similar results ($r^2 > 0.83$). One interesting note is the 16 μm size fraction often had concentrations greater than the 22.5 μm size fraction and it is not clear why this occurred, although it may be due to inaccurate pore sizes of the filters. Similar results were found between OPD and the SCOD size fractions (Figure 3-16). The results from these tests differ from the previous data in that a good correlation is shown for all the samples.

In order to compare the different samples on the same graph, Figure 3-17 displays the OPD vs the protein plus polysaccharide concentration for the 4.2 μm size fraction and with each data point labeled. The MLSS, WAS and anaerobically digested samples had the lower OPDs, followed by anaerobically digested samples and the thermophilic TPAD sample had the highest.

Table 3-4. Summary of Results from Batch Tests Examining Different Biopolymer Size Fractions. (Data Sorted by Optimum Polymer Dose).

Sample Type	Polymer Dose (g/kg)	4.2 μm Protein (mg/L)	8 μm Protein (mg/L)	16 μm Protein (mg/L)	22.5 μm Protein (mg/L)	4.2 μm Polys. (mg/L)	8 μm Polys. (mg/L)	16 μm Polys. (mg/L)	22.5 μm Polys. (mg/L)
MLSS	2.81	22	22	22	22	31.9	29.2	37.7	25.5
MLSS-Industrial	14.7	60	54	36	60	7.7	12.6	15.7	10.0
Aerobically Digested	15.3	258	345	340	330	115	125	162	151.2
Anaerobically Digested	16.3	297	445	440	450	51.7	77.3	72.8	71.4
WAS-Industrial	26.6	94	100	117	94	26.7	36.6	36.9	32.7
Anaerobically Digested	37.6	638	732	711	758	184	209	199	213
Anaerobically Digested	39.7	1243	1288	1199	1021	291	352	367	332
Anaerobically Digested	42.1	866	733	1044	999	190	245	303	302
Anaerobically Digested	43.9	600	644	600	666	342	356	390	413
TPAD	70.9	1799	2029	2622	2067	426	562	692	594

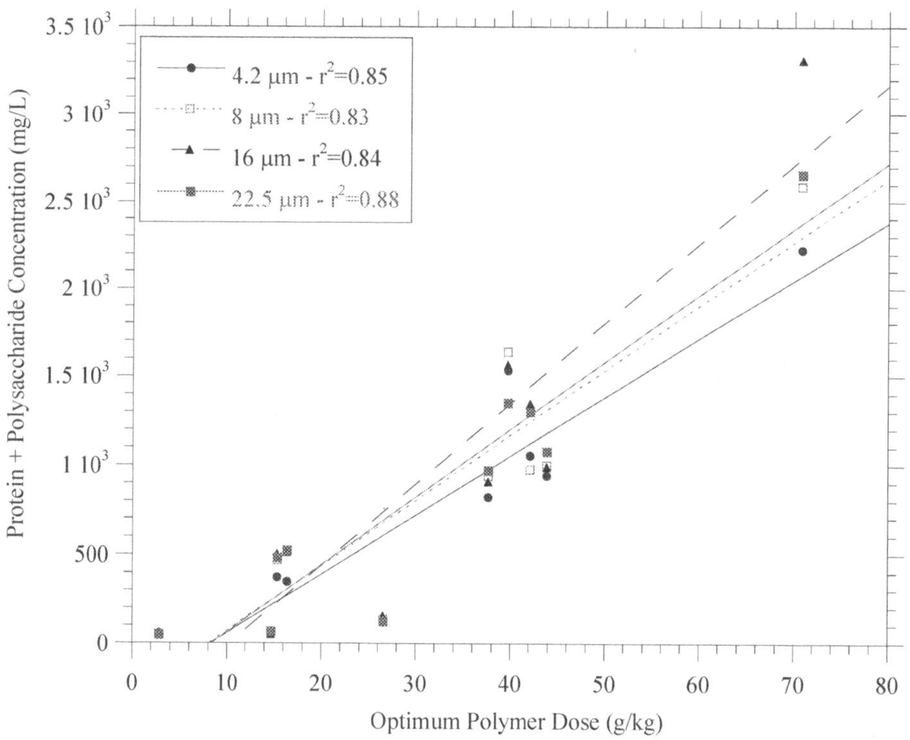

Figure 3-15. Relationship Between Different Protein and Polysaccharide Size Fractions and OPD for Different Samples.

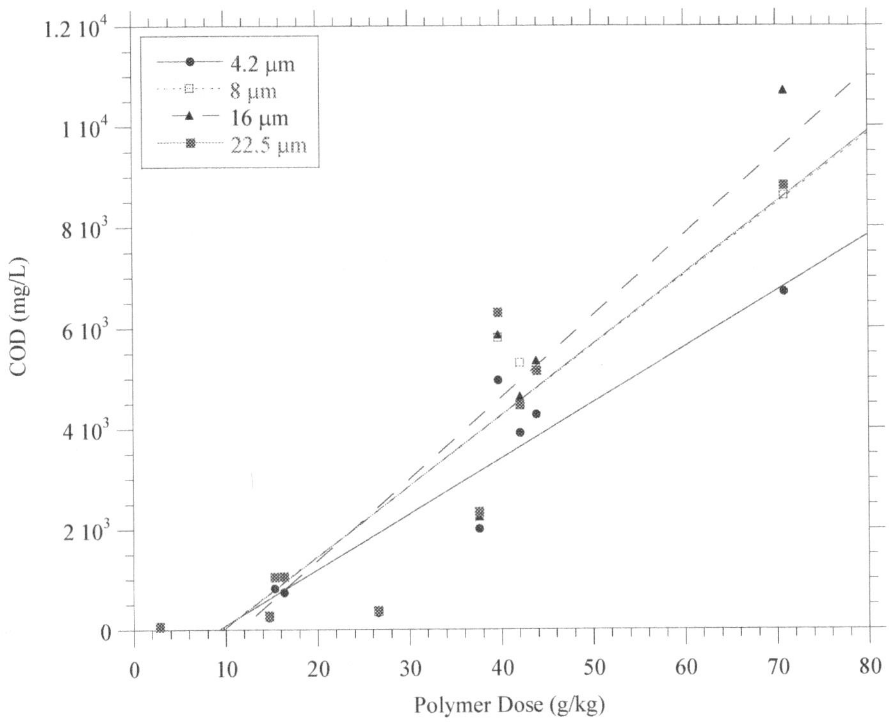

Figure 3-16. Relationship Between Different COD Size Fractions and OPD For Different Samples. (r^2 = 0.85-0.87 for all Lines).

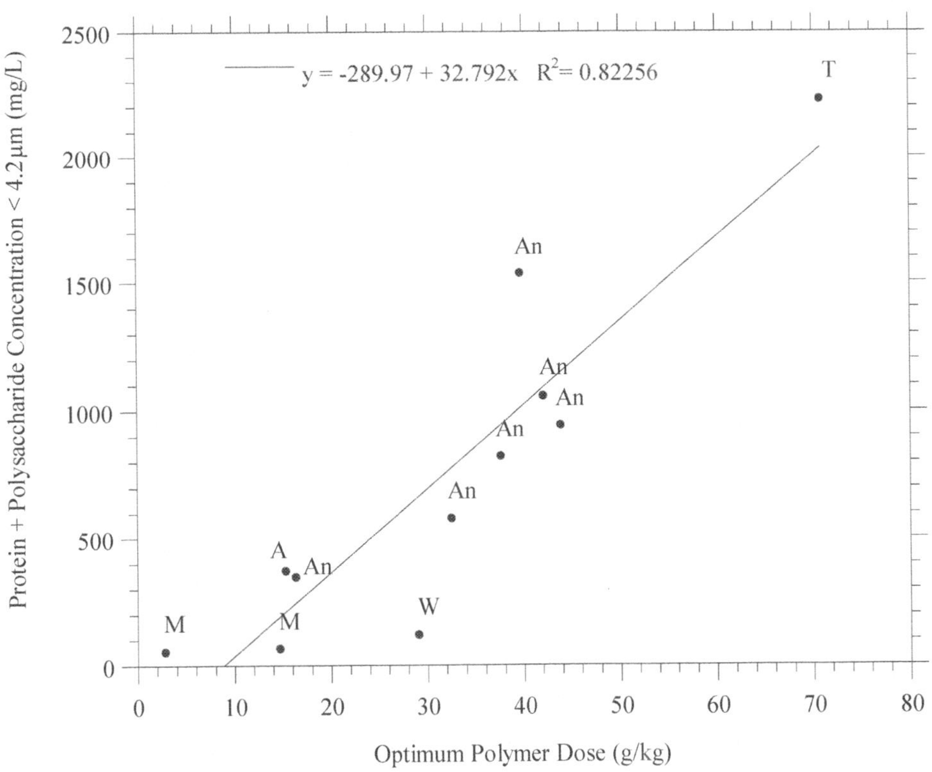

Figure 3-17. OPD vs 4.2 µm Size Fraction of Protein and Polysaccharide Showing Different Sample Types (A=Aerobically Digested, An=Anaerobically Digested, M=MLSS, T=TPAD, W=Waste Activated Sludge).

3.4 Conclusions and Implications

3.4.1 Conclusions

Based on the results from this part of the research, the following conclusions can be made:

- Biocolloids, defined as protein and polysaccharides less than 6.0 μm, create polymer demand and greater amounts of biocolloids result in greater polymer requirements for conditioning.
- Protein generally is a significantly greater fraction of the biocolloids compared to polysaccharides.
- Biocolloids have similar characteristics in terms of the amount of charged functional groups within the protein fraction.
- SCOD is a simple and representative measure for biocolloid concentration which can be used by personnel at treatment plants to understand how this parameter affects polymer demand.
- Sludges have an inherent polymer demand due to the larger, non-biocolloidal matter, and this demand appears to be similar for different types of sludges on a mass of dry solids basis.
- A weak negative correlation was found between the biocolloids and iron content of the anaerobically digested samples, and a weak positive correlation was found between the M/D ratio and biocolloid concentration.
- Greater amounts of iron in a sludge result in a decrease in the polymer demand for anaerobically digested sludges. No other metal or combination of metals appeared to be related to optimum polymer dose.
- Operational parameters such as SRT and VS destruction in mesophilic anaerobic digestion do not appear to be related to polymer demand for conditioning.
- Thermophilically digested sludges have much greater polymer demand which is associated with greater amounts of biocolloids in suspension.
- More research is needed to determine whether the relationship between biocolloids and OPD holds for additional polymer types and for high polymer demand samples..

3.4.2 Implications

These conclusions lead to several important implications for sludge management. These include:

- Decreasing the biocolloid concentration will reduce the polymer demand for conditioning. Solution phase COD or protein and polysaccharide measurements can be used to investigate factors that affect the polymer demand. For example, if a plant were investigating the impact of upgrading to thermophilic digestion and they were operating lab scale digesters, measuring the biocolloid concentration would provide information on how the polymer demand would change. This information would replace the need for actual dewatering trials which are not always practical with laboratory scale reactors.

- It may be practical to reduce the biocolloid concentration through chemical addition, especially Fe, and thereby lower the OPD. This is supported by results presented in Chapter 8.0 that demonstrate Fe addition prior to polymer addition results in sequestration of biopolymer and a significant decrease in polymer demand.

CHAPTER 4.0

EFFECT OF POLYMER CHARACTERISTICS ON POLYMER DEMAND

Understanding the stoichiometry of polymer demand requires an understanding of both the sludge constituents that create the polymer demand and the interactions of different polymer types. The research presented in Chapter 3.0 makes the case that biocolloids create a large portion of the polymer demand. However, the experiments were performed on single types of polymer. An understanding of how different polymer types interact with sludges is needed to fully understand the mechanism of polymer demand. Polymer development has focused on increasing the molecular weight of the polymer to improve the dewatering properties. More recently, polymer manufacturers are developing structured polymers that are branched or cross-linked (Pearse et al., 2001). In addition, polymers are available with hydrophobic moieties that may enhance dewatering.

4.1 Effect of Polymer Charge and Configuration

The effect of different polymer characteristics on polymer demand was investigated using 14 different polymers. This initial investigation examined polymers with different configurations, charge, and functional groups (Table 4-1). The charge density of the polymers was determined using a PVSK titration and toludine blue indicator (Kam and Gregory, 1999; Dentel et al., 1993).

Table 4-1. Summary of Different Polymer Characteristics.

Supplier	Polymer Name	Mole Charge	Structure and Molecular Weight	Charge Density (meq/g)	Std. Dev. of Charge Titration (meq/L)
Betz	CE2682	15	Linear, MW 6-9 million	0.401	0.033
Betz	CE2680	40	Linear, MW 6-9 million	1.061	0.023
Betz	CE2688	80	Linear, MW 6-9 million	1.924	0.085
Betz	CE2672	5	Branched, MW 5-8 million	0.164	0.006
Betz	CE2676	20	Branched, MW 5-8 million	0.605	0.019
Betz	CE2653	40	Branched, MW 5-8 million	1.080	0.022
Betz	CE2660	20	Linear, Hydrophobic, MW = 7-10 mil.	0.591	0.011
Betz	CE2665	35	Linear, Hydrophobic, MW = 7-10 mil.	0.771	0.020
Betz	CE2667	55	Linear, Hydrophobic, MW = 7-10 mil.	1.312	0.024
Cytec	PWD	ND	Unknown	0.831	0.001
NALCO	7194	50	Linear	0.717	0.013

In the first study, three polymers were used to condition the same sludge sample. The three polymers had the same molecular weight and chemistry, and varied only in their charge density (Table 4-2).

Table 4-2. Summary of Polymer Characteristics Used in Conditioning Study.

Configuration	Molecular Weight	Manufacturer	Trade Name	Mole Charge (%)
Linear	6-9 x 10^6	Betz	CE2681	10
Linear	6-9 x 10^6	Betz	CE2680	40
Linear	6-9 x 10^6	Betz	CE2688	80

A sludge sample was collected from an anaerobic digester and the CST-polymer dose profile was determined for each sample. The OPD was determined and the results for the three polymers are plotted in Figure 4-1. The OPD is plotted using the typical units of grams of polymer per kg of dry solids (g/kg DS) and also as equivalents of charge per kg of dry solids (meq/kg DS). The two units show opposite trends. As the charge density of the polymer increased, the polymer dose expressed on a weight basis decreased from approximately 32 g/kg to 13 g/kg. This would be expected since each gram of polymer would have greater numbers of charge to neutralize the anionic biocolloids, and therefore less mass of polymer would be required. However, when the polymer dose is expressed on a charge basis (using the charge density of the polymer), the opposite trend is found. The lower the charge density, the less polymer charge is used in conditioning the sample. Therefore, the lower charge density polymers use their charge more efficiently compared to the higher charge density polymers.

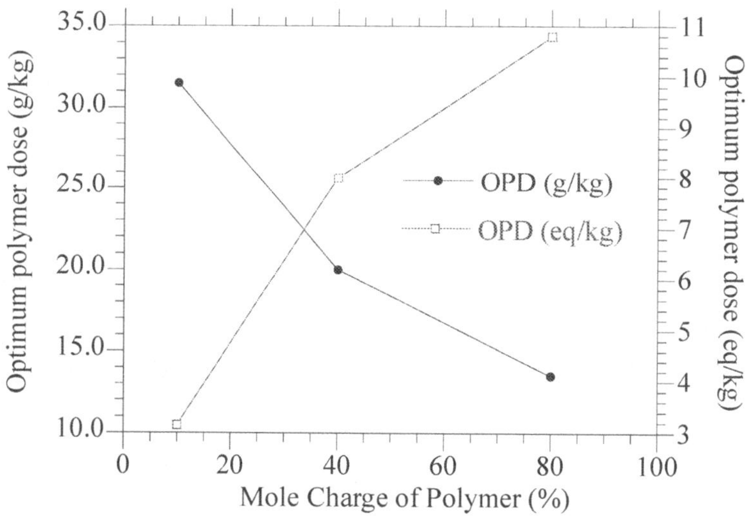

Figure 4-1. Optimum Polymer Dose Expressed as g/kg and meq/Kg for Polymers with Varying Charge Density.

This phenomenon may be explained by considering the interactions of polymer with particles. A typical model for polymer adsorption assumes that portions of the polymer adsorb to the particle surface and other portions form loops and tails that do not adsorb (Gregory, 1993). These loops and tails would potentially contain cationic charges that do not result in charge neutralization (assuming they do not bridge to other particles). Therefore, the more charge per length of polymer, the greater the amount of unused cationic charge. Figure 4-2 shows a simplified schematic of this phenomenon.

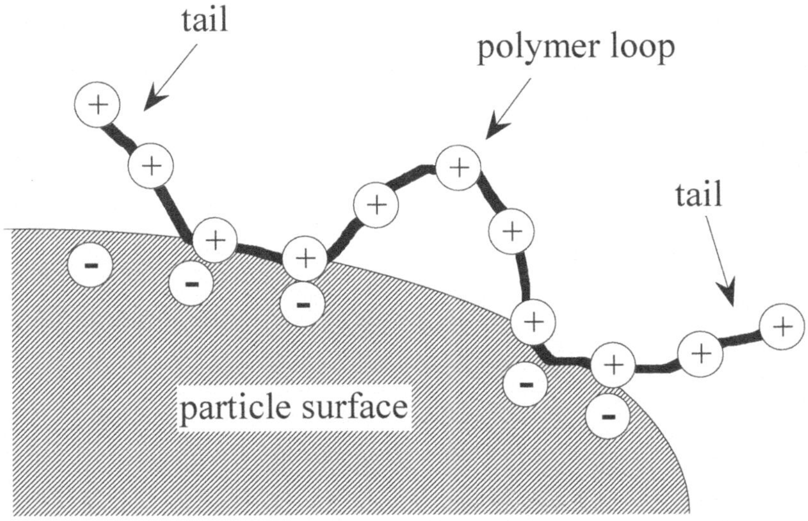

Figure 4-2. Schematic of Polymer Loops and Tails Showing Unused Charge of Polymer.

To further confirm the biocolloid theory, the solution phase (< 6.0 μm) protein and polysaccharide were measured during a set of polymer conditioning experiments. For each of the polymer samples, the addition of increasing amounts of polymer resulted in decreased solution protein (Figure 4-3) and polysaccharide (Figure 4-4). More protein was removed (about 400 mg/L) than polysaccharide (about 70 mg/L), but when expressed as a percentage of the original amount, about 83% of protein and 80% of polysaccharide was removed from solution. These data support the biocolloid theory since the biocolloids are removed from solution at increasing polymer dosages. It is interesting to note that the protein and polysaccharide concentrations continue to decrease even as more polymer is added beyond the optimum polymer dose. This could have implications for odor production since odor production from sludge has been shown to be related to sludge cake protein (Higgins et al., 2005a).

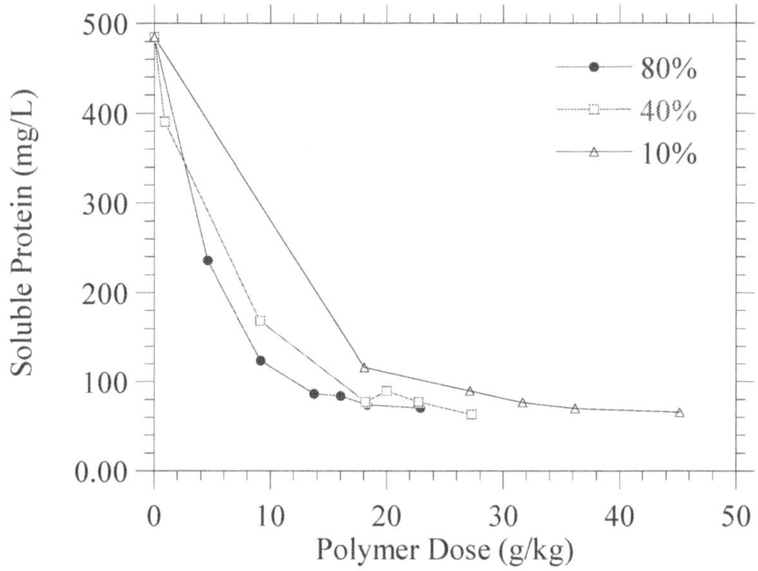

Figure 4-3. Effect of Polymer Addition on Soluble Protein Concentration Remaining in Solution.

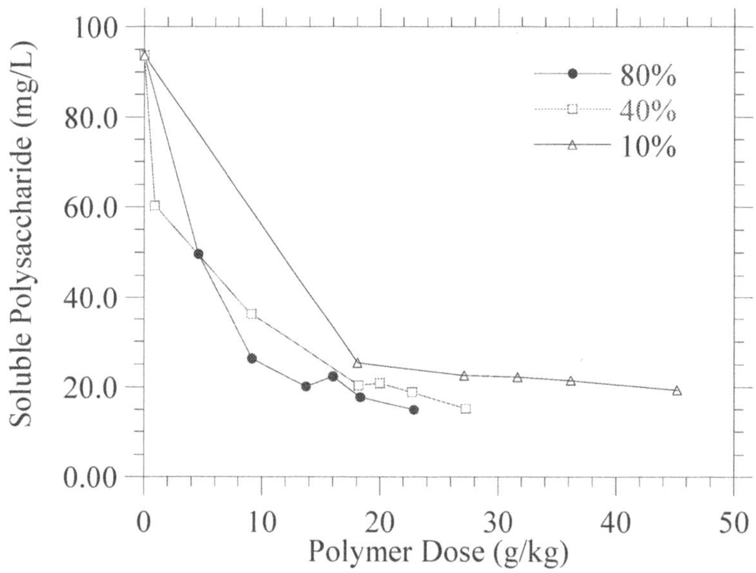

Figure 4-4. Effect of Polymer Addition on Soluble Polysaccharide Concentration Remaining in Solution.

A second, more extensive set of experiments was performed using anaerobically digested sludge, and nine different polymers. The polymers used were supplied by GE-Betz. Polymers with different charge densities were used from each different polymer type: four linear (L); two branched (B); and three hydrophobic (H). The OPD on a mass of polymer per mass of dry solids basis as a function of the polymer charge density is plotted in Figure 4-5. The data show no clear relationship for polymer demand, and polymer configuration (e.g., linear, branched, or hydrophobic) did not appear to have an effect on OPD. The relationship between OPD on an equivalent basis and the polymer charge density is shown in Figure 4-6. These results clearly show that the polymer charge density is the key parameter in determining polymer dose and that polymer demand should be expressed on an equivalent charge basis to normalize for different polymers. The greater the charge density, the greater the polymer demand. This result is consistent with the model illustrated in Figure 4-2; more highly charged polymers have reduced efficiencies in the use of their charge to neutralize the negatively charged particles.

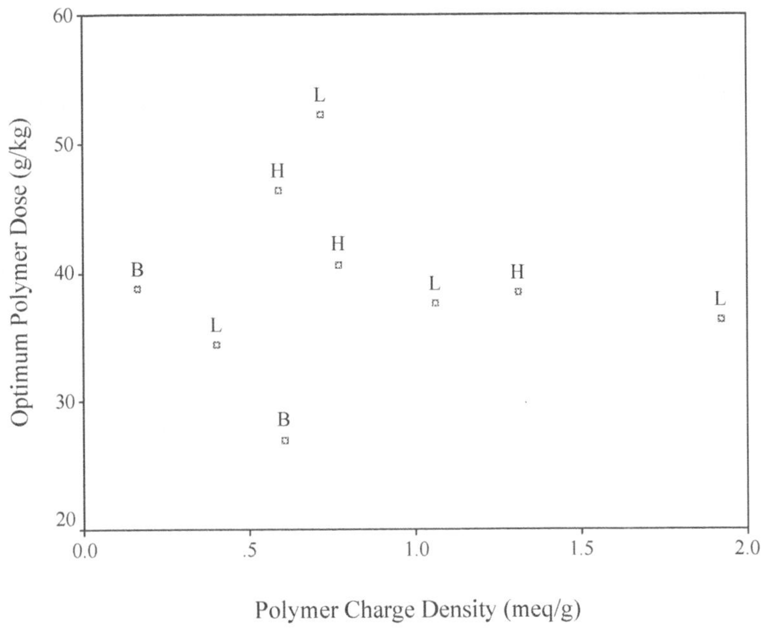

Figure 4-5. Relationship Between OPD on a Mass Basis and Polymer Charge Density for Different Polymer Configurations (L=Linear, B=Branched, H=Hydrophobic).

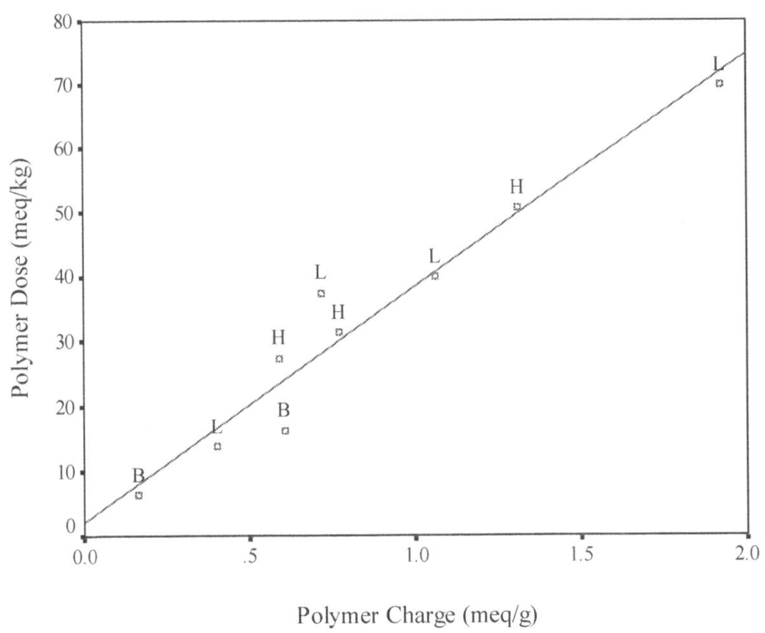

Figure 4-6. Relationship Between OPD on an Equivalent Basis and Polymer Charge Density for Different Polymer Configurations (L=Linear, B=Branched, H=Hydrophobic).

The protein, polysaccharide and SCOD remaining in solution at the optimum dose were measured for the different size fractions (4.2, 8, 16, and 22.5 μm) to determine if the different polymer types and/or charge densities had different biopolymer removal efficiencies. The results were similar for the different size fractions. A plot of the 8 μm size fraction of protein plus polysaccharide concentration remaining in solution at the OPD as a function of the polymer charge is shown in Figure 4-7. The results show that the greater the polymer charge, the greater the efficiency in removing the solution phase biopolymer. The correlation coefficient for the linear regression is 0.63. These results fit well with the OPD data in that higher charge densities result in greater polymer demand on an equivalent charge basis, and higher charge densities also remove greater amounts of solution biopolymer. Therefore, polymer with higher charge densities can deposit more biopolymer into the cake, which could impact return stream quality and formation of odor causing compounds in the cake (Higgins et al., 2005b). The linear and hydrophobic polymers performed similarly, while the branched polymers were the least efficient in removing the biopolymer from solution. A similar plot is shown in Figure 4-8 for the relationship between protein and polysaccharide and the OPD. Greater amounts of polymer at the OPD resulted in greater biopolymer removal. Figure 4-9 shows that greater polymer charge also results in lower CST values at the OPD.

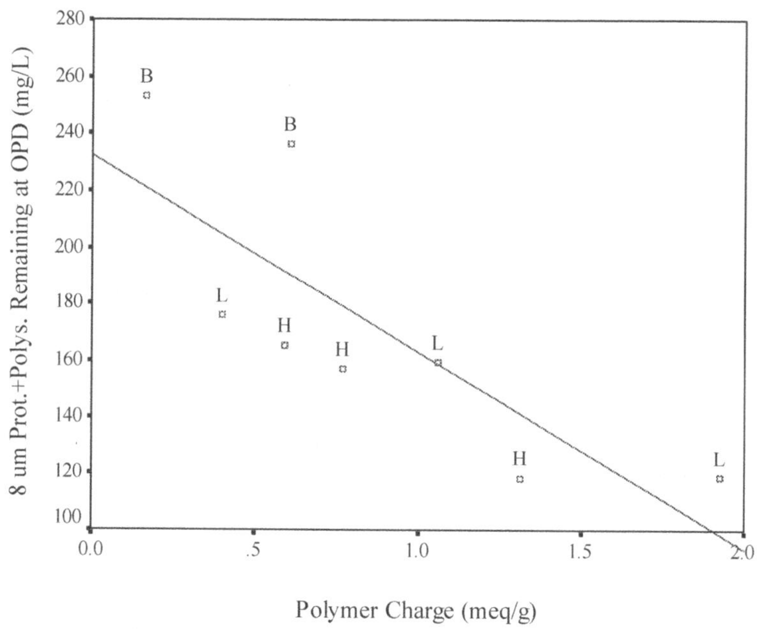

Figure 4-7. Protein Plus Polysaccharide Concentration Remaining in Solution at the OPD as a Function of the Polymer Charge Density, Size Fraction of < 8 μm (L=Linear, B=Branched, H=Hydrophobic).

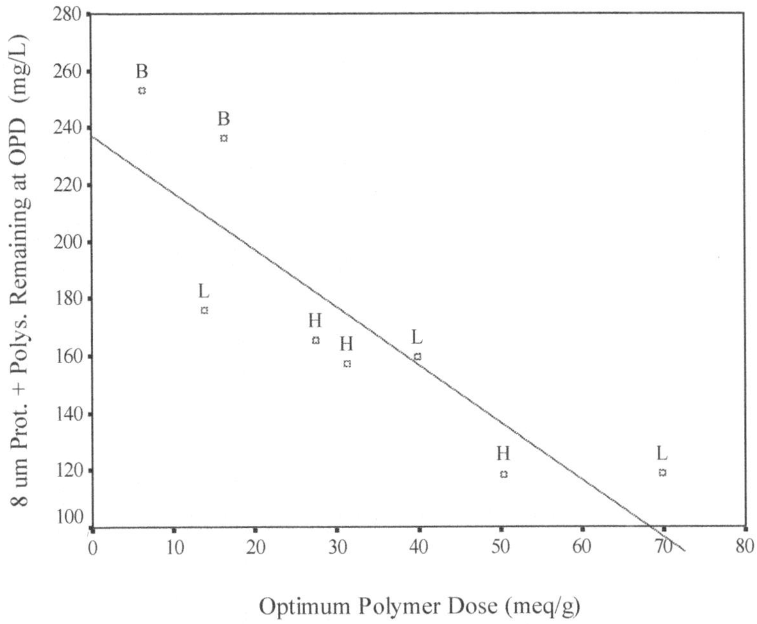

Figure 4-8. Protein Plus Polysaccharide Concentration Remaining in Solution at the OPD as a Function of the OPD (Size Fraction Of < 8 μm).

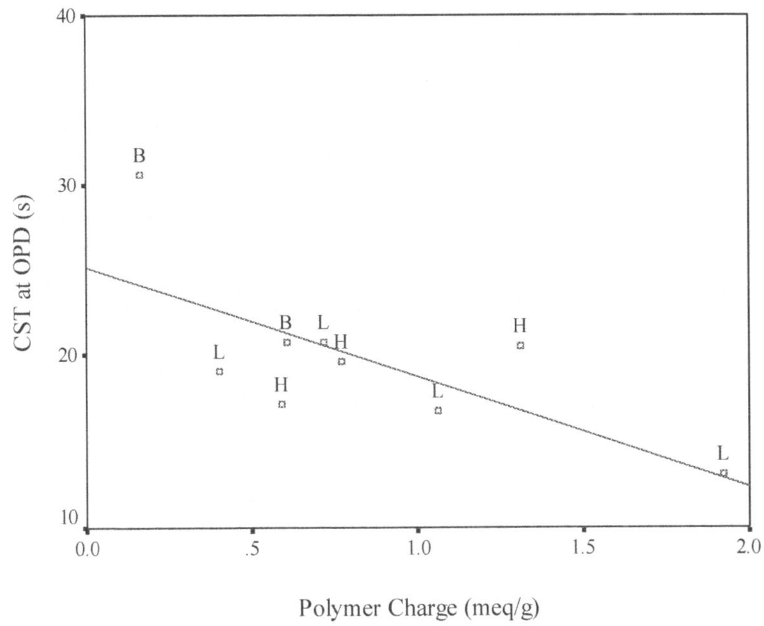

Figure 4-9. CST at the OPD as a Function of the Polymer Charge Density.

The results from these experiments show that charge density of the polymer is a key parameter for determining polymer demand for conditioning. An important caveat for these experiments is they were all performed at the same shear, as measured by the Gt value of approximately 20,000 (method of calibration is provided below). The experiments do not show how the different polymers would behave when different amounts of shear are imparted to the sludge. Shear has been shown to be a key component affecting polymer demand (Novak and Lynch, 1990; Novak et al., 1988; Werle et al., 1984). Chapter 5.0 examines the impact of shear on polymer demand by different polymers.

4.2 Effect of Polymer Molecular Weight

The role of polymer molecular weight (MW) was studied for a series of polymers with MWs that ranged from 25,000 to 11 million, and with different chemistries and charge densities. The charge densities were all determined using the PVSK titration method as described previously. A summary of the polymers used in this study are provided in Table 4-3.

Table 4-3. Summary of Polymer Characteristics for Effect of MW Study (All Cytec Polymers).

Polymer Name	Mole Charge	Other Characteristics	Molecular Weight	Solution Type	Charge Density (meq/g)	S.D. of Charge Titration (meq/g)
C-572	100	Polyamine	25,000	emulsion	3.571	0.015
C-577	100	Polyamine	100,000	emulsion	3.430	0.012
C-581	100	Polyamine	250,000	emulsion	3.387	0.043
C-587	100	POLY-DADMAC	60,000	emulsion	1.204	0.017
C-591	100	POLY-DADMAC	400,000	emulsion	1.258	0.022
C-595	100	POLY-DADMAC	900,000	emulsion	2.395	0.007
C-1592	10	Resin Amine	8 million	emulsion	0.379	0.027
C-1594	20	Resin Amine	8 million	emulsion	0.665	0.029
C-1596	40	Resin Amine	8 million	emulsion	1.118	0.047
MX-20	10	Polyacrylamide	11 million	emulsion	1.300	0.030
MX-40	20	Polyacrylamide	11 million	emulsion	0.730	0.022
MX-60	40	Polyacrylamide	11 million	emulsion	1.065	0.063
SD2065	35	Branched	3 million	emulsion	1.254	0.030
SD2061	40	Branched	3 million	emulsion	1.047	0.028
SD2080	55	Branched	3 million	emulsion	1.450	0.039
SD2085	55	Branched	3 million	emulsion	1.190	0.055

For these experiments, the OPD was found using 15 or 16 of the polymers and three different sludge types (anaerobically digested, aerobically digested, and waste activated). The effect of MW on the OPD for the three different sludge types is shown in Figures 4-10, 11, 12. The OPD varied considerably on a charge basis for all three sample types. Since charge neutralization is often cited as the mechanism for conditioning, the effect of MW would not be expected, especially when plotting the OPD in terms of charge equivalents of polymer added per kg of sludge. However, the results clearly show higher MW polymers require less overall charge for conditioning. Although the actual amounts for conditioning were different for the three sludge types, the overall trends were the same; higher MW polymers required less charge to reach the optimum dose. It is possible that lower MW polymers have greater access to the negatively charge sites within certain areas of the flocs, whereas larger MW polymers are sterically inhibited from binding within these less available sites. As a result, less overall charge is needed for neutralization. The results show that MWs greater than about 1-2 million do not result in any significant additional decrease in polymer demand. An additional explanation is the role of polymer bridging in conditioning. Lower MW polymers would be less able to bridge the biocolloids compared to higher MW polymers. Therefore, bridging is likely a key mechanism for floc formation, as well as for resistance to shear during dewatering. The resistance to shear and the role of polymer bridging is presented more fully in Chapter 5.0.

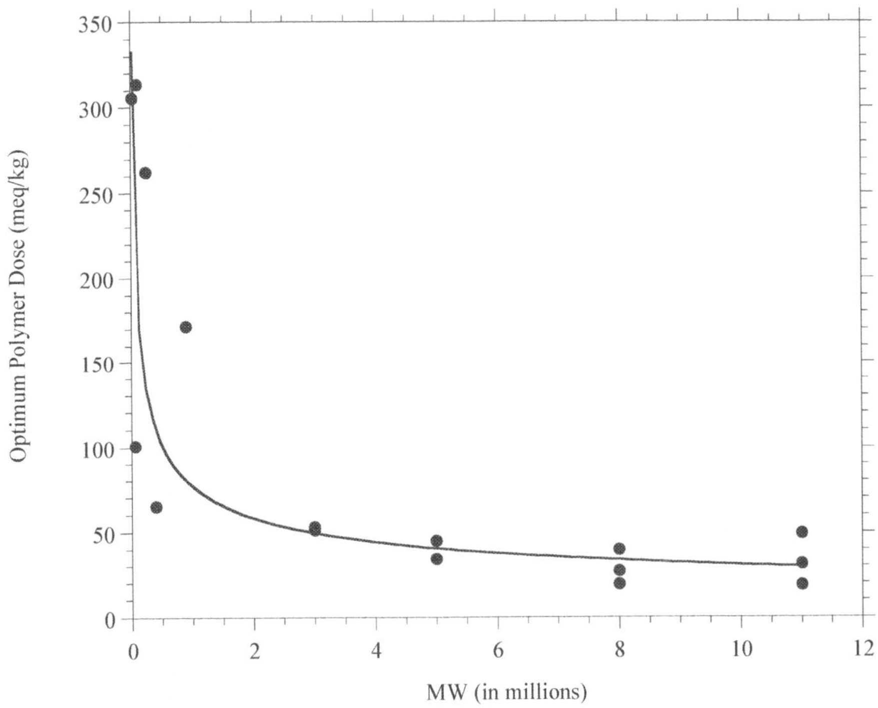

Figure 4-10. OPD for Anaerobically Digested Sludge for Polymer Set with Varying Molecular Weight.

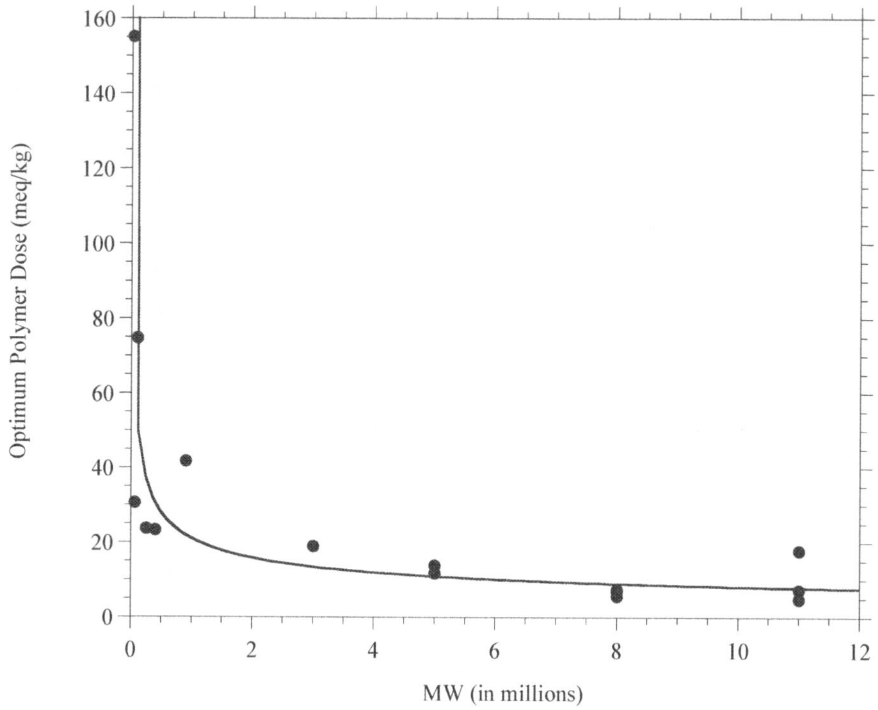

Figure 4-11. OPD for Aerobically Digested Sludge for Polymer Set with Varying Molecular Weight.

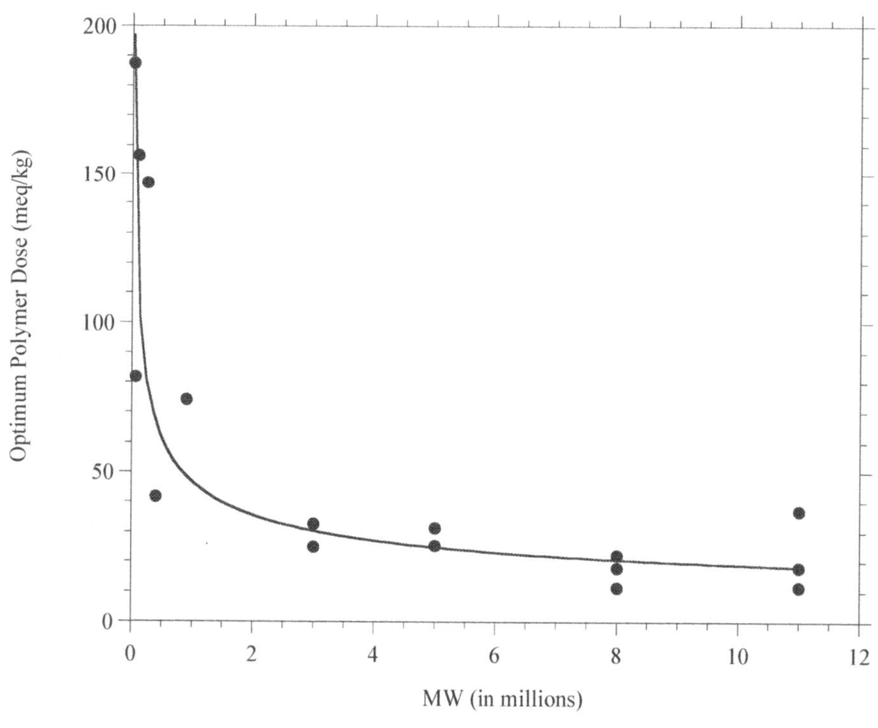

Figure 4-12. OPD for Industrial Waste Activated Sludge for Polymer Set with Varying Molecular Weight.

A comparison of SCOD (4.2 μm) and the CST at the OPD for the different MW polymers and the anaerobically digested sample is shown in Figure 4-13. There was no significant difference in SCOD at the OPD for the range of MWs examined, although the general trend was more SCOD left in solution at the OPD for higher MW polymers and for all three types of sludges. However, the CST at the OPD decreased with the higher MW polymers. This suggests that the conditioned product produced by the higher MW polymers is a better dewatering sludge than when low MW polymers were used.

In general, the same patterns were seen for the aerobically digested and waste activated sludges (results not shown). However, a greater amount of scatter was measured for the SCOD with the aerobically digested sample.

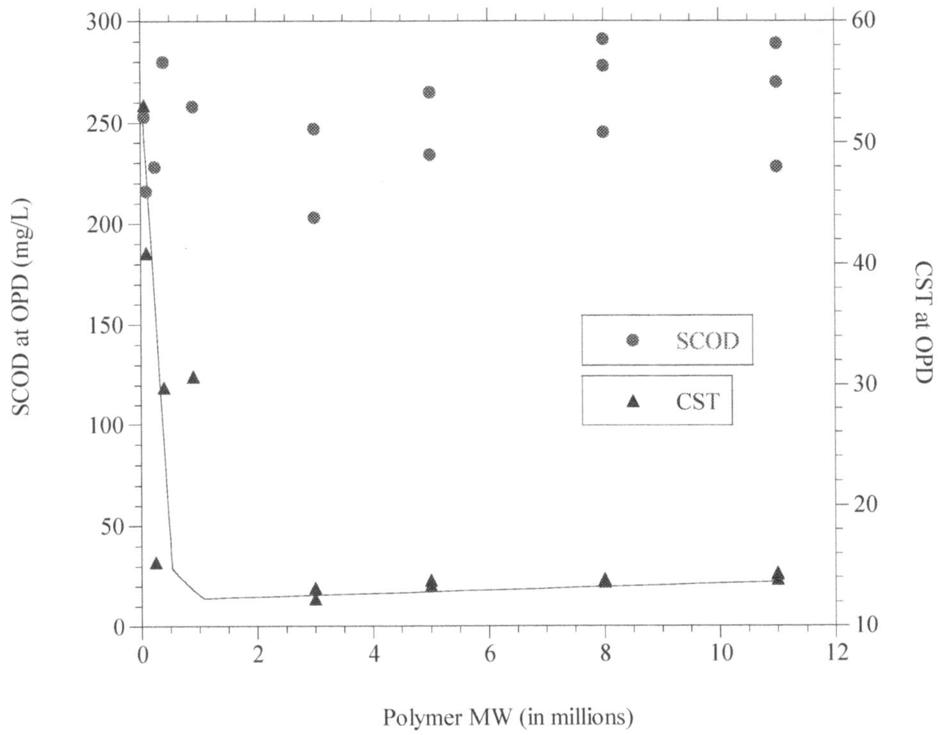

Figure 4-13. SCOD and CST at the OPD for the Different MW Polymers and Anaerobically Digested Sample.

4.3 Conclusions and Implications

4.3.1 Conclusions

Based on the results from this part of the research, the following conclusions can be made:

- The polymer charge density and molecular weight are key parameters in determining the OPD for a given sludge.
- For high MW polymers (greater than about 1 million), the greater the charge density of the polymer, the lower the polymer dose on a mass basis. However, on a charge basis, the polymer dose increased linearly with increasing polymer charge density, suggesting a less

efficient use of charge. The higher charge density polymer typically produced better dewatering characteristics at the OPD.

- As the molecular weight of the polymer increased, the polymer dose decreased significantly. Increasing MW above 1 million did not have a significant impact on polymer dose. Low MW polymers (< 1 million) had significantly greater polymer demand, and the sludge floc at the OPD had worse dewatering properties compared to the higher MW polymers.
- The results demonstrate that polymer bridging is an important mechanism for conditioning in addition to charge neutralization. Floc formation requires bridging by polymer for strength to withstand shear associated with dewatering.

4.3.2 Implications

Based on these results, several implications for practice are evident.

- When choosing polymers for dewatering, molecular weights greater than 1 million should be considered for best dewatering performance. Low MW polymers likely will not provide flocs with the strength to withstand shear during dewatering.
- Higher charge density polymers will provide better dewaterability and improved centrate or filtrate quality since they are able to better capture the solution-phase biocolloids and deposit them within the cake. However, other research has shown that greater amounts of protein deposited in the cake during dewatering results in increased odor production during cake storage. Research is needed to better understand the role of polymer characteristics on product quality, such as odor characteristics of the cake.

CHAPTER 5.0

IMPACT OF SHEAR ON POLYMER DEMAND

5.1 Calibration of Mixer

The shear or mixing intensity imparted to a suspension is typically quantified using the velocity gradient, G (s^{-1}) multiplied by the time of mixing, t. This unitless parameter, or Gt value, can be used to examine the impact of shear on sludge conditioning (Novak and Lynch, 1990; Novak et al., 1988; Werle et al., 1984). The G value for a given mixer setup (paddle, reactor vessel, and speed) can be determined by creating a curve of G versus mixer speed for the mixing regime. Typically, torque is measured and from this, G is calculated (see Werle et al., 1984). The velocity gradient as a function of mixer rpm was determined for the mixer setup at Bucknell University using a torque meter that was calibrated with a torque wrench. The mixer setup used a 2-L, baffled mixing vessel with 1 L of sample and a flat paddle blade. The calibration curve for the mixer is shown in Figure 5-1.

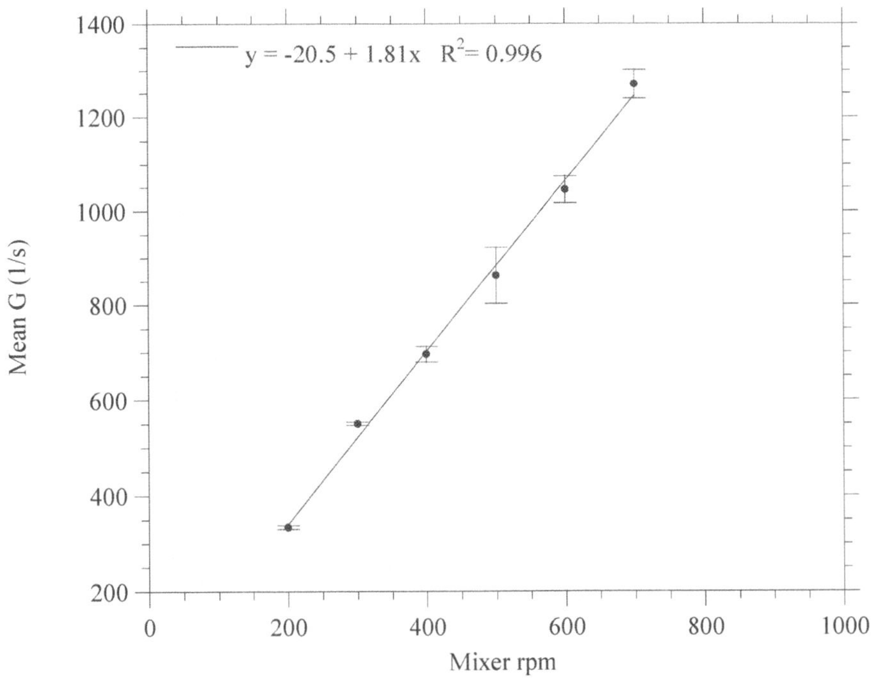

Figure 5-1. Calibration Curve for Determining Velocity Gradient (G) as a Function of Mixer rpm.

5.2 Impact of Shear on Polymer Demand

The impact of shear can be determined by mixing sludge samples at a constant intensity, measured by G, with different polymer doses for different time intervals. CST is then measured at the different time intervals to develop a family of Gt curves. For example, Figure 5-2 shows the CST as a function of Gt for three different polymer types at the same polymer dose (18 meq/kg). The sample was an aerobically digested biosolids.

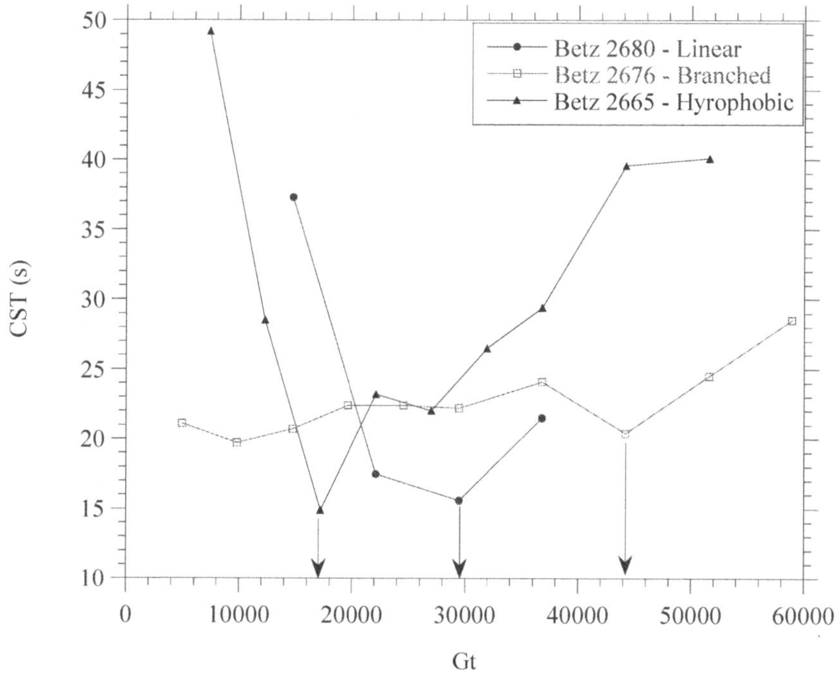

Figure 5-2. Gt vs CST for Similar Polymer Dose on an Equivalent Basis (18 meq/kg DS) for Three Different Types of Polymers.

Using theis family of curves, the Gt that the corresponding polymer dose can withstand typically results in the minimum CST. For example, in Figure 5-2 the arrows indicate the Gt value of minimum CST for the given polymer and polymer dose. As the Gt increased beyond this point, the dewatering properties deteriorated. This point then provides a polymer dose-Gt data point (i.e., the OPD for that Gt value). Using this information, OPD versus Gt curves can be developed to examine how polymers and sludge types may respond to shear. For example, a series of shear experiments were performed using three different polymers from previous experiments with a range of Gt values from approximately 5,000 - 70,000. The polymer dose on a mass basis as a function of the Gt with the minimum CST is shown in Figure 5-3. The results from these experiments show that for a given dose, the linear and branched polymers have similarly shaped curves, although the magnitude of the OPD is different, with the branched polymer requiring a greater mass of polymer.

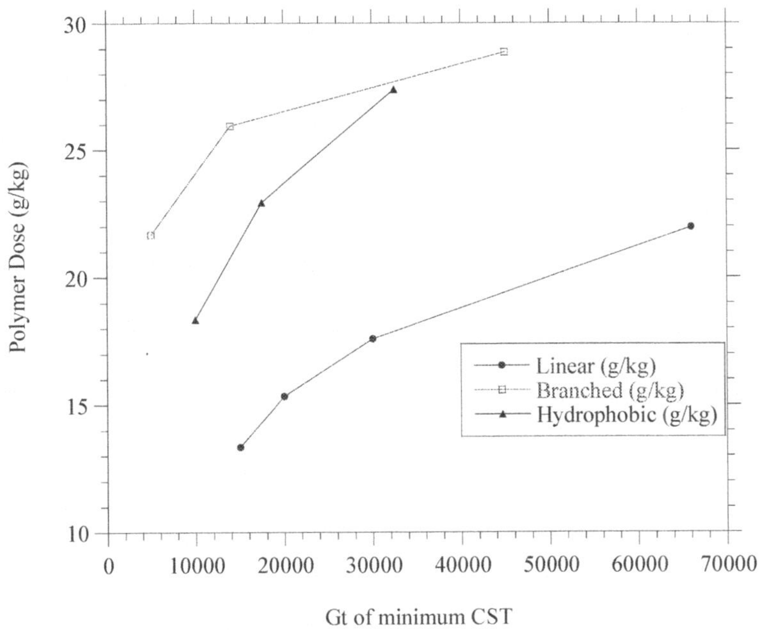

Figure 5-3. Polymer Dose as a Function of Gt Values Giving the Minimum CST.

When these data are plotted using polymer dose on an equivalent basis, less of a difference is seen in the polymer demand (Figure 5-4). The branched polymer appears to perform better at higher shear, in that the increase in polymer dose as a function of Gt begins to level off at the higher Gt values. In comparison, the linear and hydrophobic polymers continue to increase almost proportionally with the Gt. This example demonstrates how this approach can be used to study how different solids/digestion and polymer types respond to shear. These experiments are discussed in the following sections.

For this data set, the higher polymer doses at the higher Gt values resulted in lower CST values (Figure 5-5). This indicated better dewaterability at higher shear, even though the polymer dose is greater. The impact of shear on resultant cake solids is investigated further in Section 5.5.

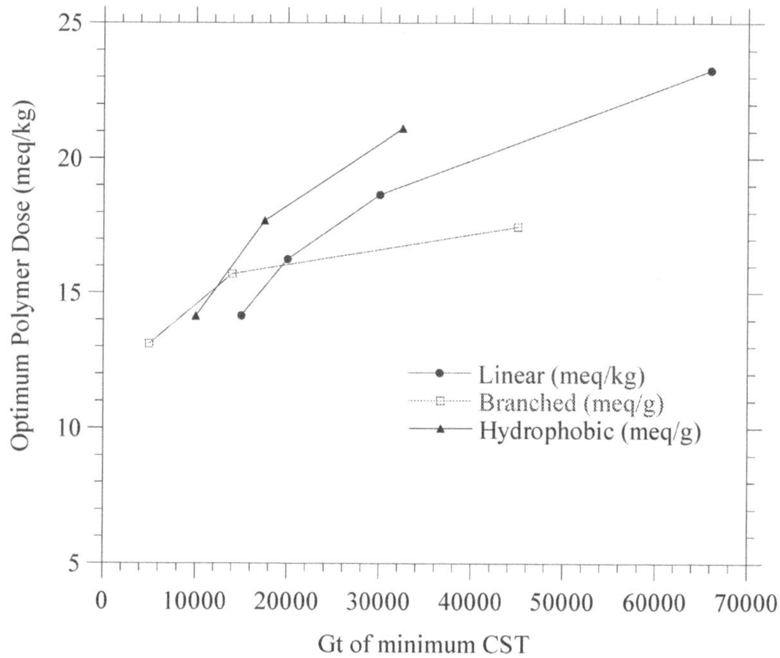

Figure 5-4. Gt of Minimum CST for Three Different Polymers at Different Dosage on a Charge Basis.

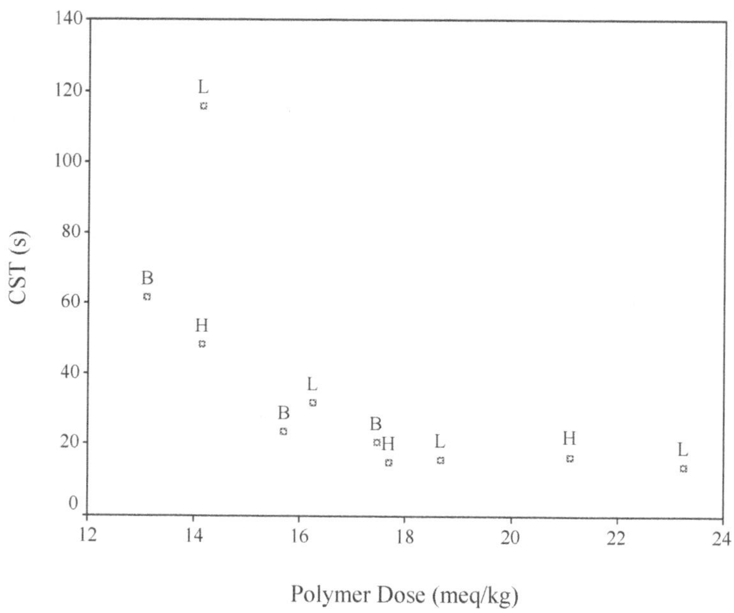

Figure 5-5. Relationship Between CST and Polymer Dose for the Different Shear Experiments (H=hydrophobic, L=linear, B=branched).

5.3 Comparison of Different Sludge Types Response to Shear

A set of experiments was performed to examine how different sludge types respond to shear. Different sludge types were collected and dewatered using the same polymer type, but with varying Gt values to developed the OPD versus Gt curves. Five sludge types—anaerobically digested, aerobically digested, thermophilic anaerobic, temperature phased anaerobic digestion

(TPAD), and waste activated (WAS)—were tested to examine their response to shear. A comparison of the polymer dose versus Gt curves for the four different digested sample types is shown in Figure 5-6. The results indicated that conventional aerobic and anaerobically digested sludges respond similarly to shear, as shown by the similar slopes and intercept of the line. However, the slopes increased for the TPAD and thermophilic sample. The WAS sample was the most sensitive to shear, as shown by the greatest slope of the OPD vs Gt curve.

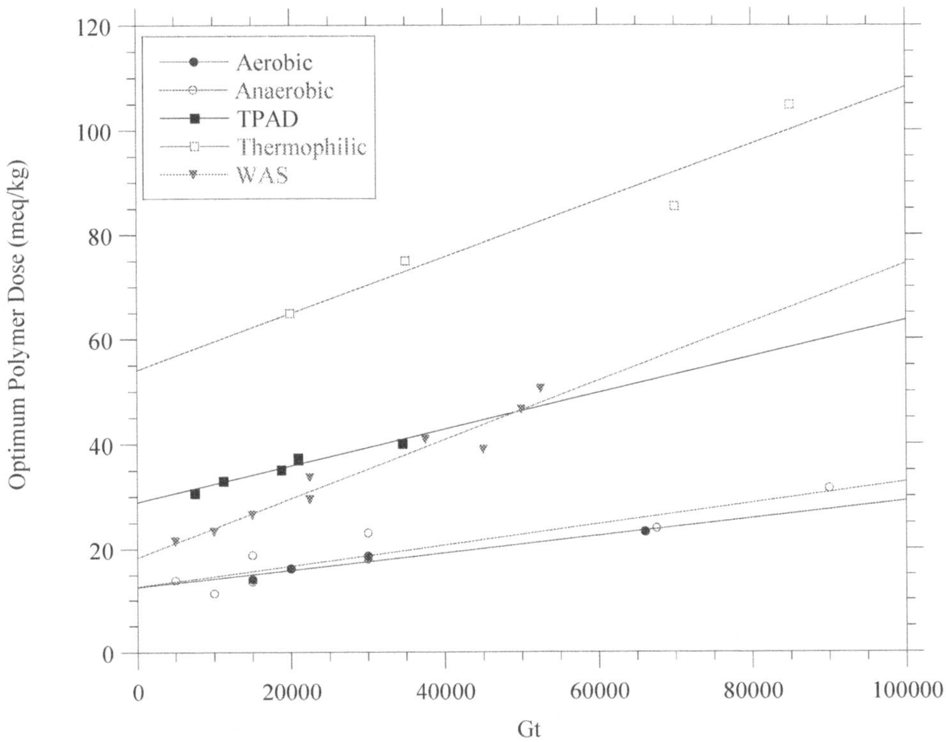

Figure 5-6. Comparison of Shear Response for Four Different Digested Sludge Types Using the Same Linear Polymer.

A summary of the slopes and intercepts for the curves in Figures 5-6 are provided in Table 5-1. The table also contains another anaerobic sample not included in Figure 5-6 for clarity, but this line fit very well with the aerobic and anaerobic samples on the graph. The results from this study show that shear susceptibility is variable depending on the digestion process used, and the order from lowest to highest is: aerobic ≈ anaearobic < TPAD < thermophilic ≈ WAS. Additional sampling is needed to further explore the relationship between sludge type and shear sensitivity, and determine if these relative relationships hold true for more samples. The results suggest for digested samples, the greater the temperature of digestion (time-temperature) results in less inherent floc strength. However, WAS with no digestion also results in relatively weak flocs. The shear susceptibility of the digested samples was associated with an increase in the 'no shear' polymer dose, taken as the y-intercept, and this also was linked to the solution biocolloids as measured by the protein and polysaccharide concentration. The higher temperature digestion appears to cause greater floc breakup and release of biocolloids, and the remaining particles also are more susceptible to breakup during shear.

Conceptually, the y-intercept can be thought of as the polymer or charge needed to neutralize both the biocolloidal fraction and larger particles, since this case represents no shear. The increase in polymer demand during shear is a result of increased availability of charged sites that become exposed during floc breakup. The smaller particles that form will need to be incorporated back into the larger floc by polymer bridging mechanisms, and this additional demand associated with shear can be thought of as demand for polymer bridging.

Table 5-1. Summary of Gt vs OPD Slopes and Intercept Data.

Sample	OPD-Gt Slope	y-intercept (OPD at no shear) (meq/kg)	Initial Biocolloid Concentration (mg/L)
Aerobic	0.00017	12.6	373
Anaerobic (on graph)	0.00020	12.7	820
Anaerobic (not shown on graph)	0.00016	17.6	no data
TPAD	0.00035	29.0	2230
Thermophilic	0.00054	54.1	3960
WAS	0.00056	18.4	121

5.4 Impact of Different G Values on Shear Response

The previous results describe the effect of shear (as measured by Gt) on the OPD for different sludge and polymer types. The experiments were all performed using the same G-value of 500 s^{-1}. Researchers have examined how different sludge types respond to different G-values (Werle et al., 1984; Novak and Bandak, 1989), but not all the sludge types investigated in this study have been examined for their response to G-value, nor have different polymer types been examined. Therefore, experiments were performed to examine the effect of different G-values (500, 1000, and 1500 s^{-1}) and Gt on the OPD using two different polymer types (branched and linear) and two different solids types (anaerobically digested and WAS).

The effect of G on the OPD for a linear polymer and anaerobically digested biosolids is shown in Figure 5-7. In general, the different G-values did not result in a different response when comparing Gt versus the polymer dose curve as shown by the linear trend for all the G-values and polymer dose combinations in Figure 5-7. Similar results were found for the branched polymer as shown in Figure 5-8. Although a comparison of the two polymers (Figure 5-9) indicates that the two polymer types may fit a similar trend, there are indications that the branched polymer may have a greater increase in polymer dose as G increases when compared to the linear polymer. For example, the slopes of the curves in Figures 5-7 and 5-8 reveal that the branched polymer dose increased at a higher rate than the linear polymer. However, the Gt range used in the branched polymer experiments were lower than the linear polymer. Additional studies are needed to examine a larger Gt range for the branched polymer to provide a better comparison of the different polymer types.

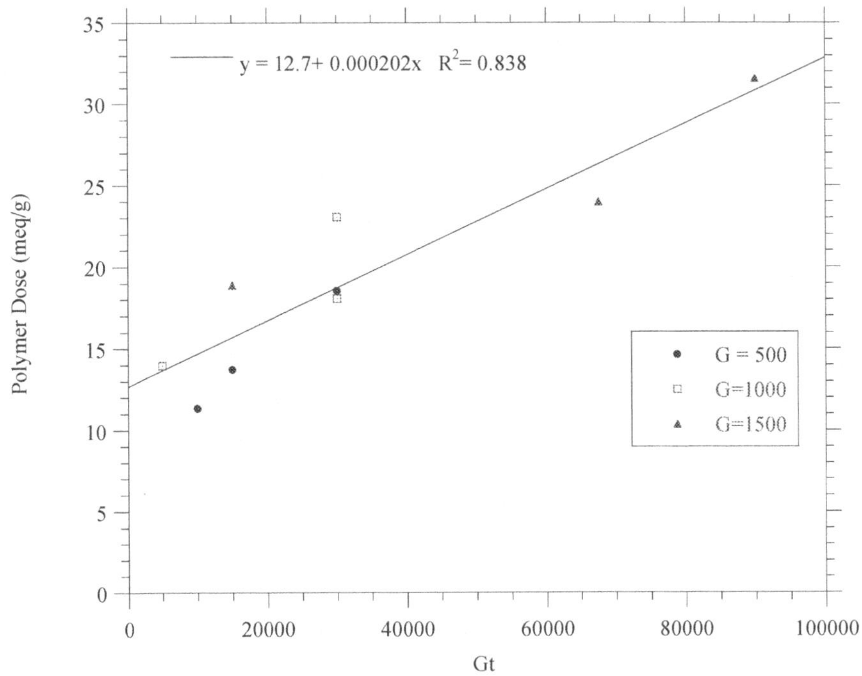

Figure 5-7. Effect of G on OPD for Linear Polymer and Anaerobically Digested Sludge.

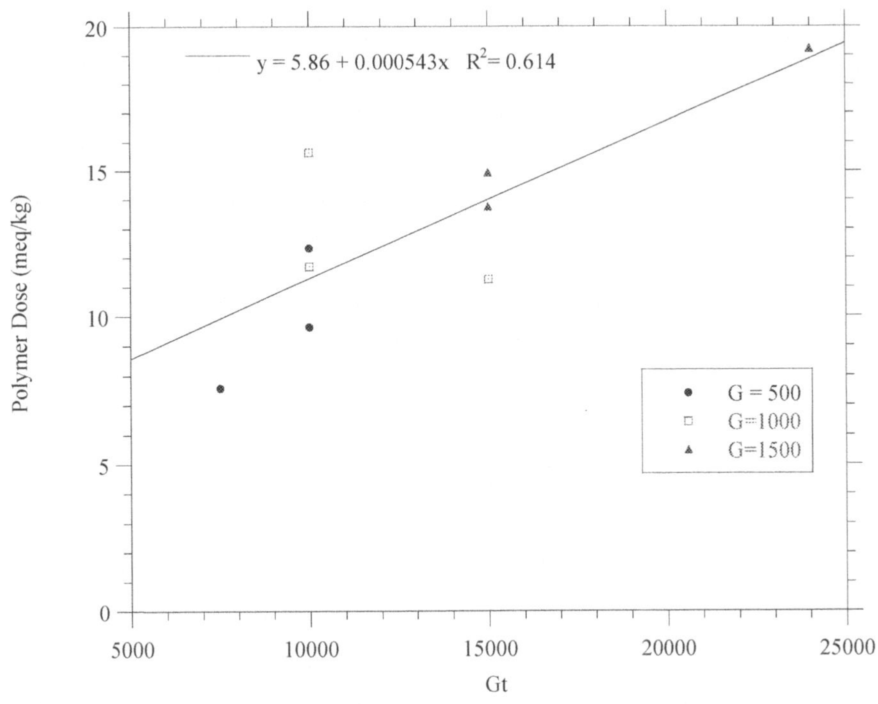

Figure 5-8. Effect of G on OPD for Branched Polymer and Anaerobically Digested Sludge.

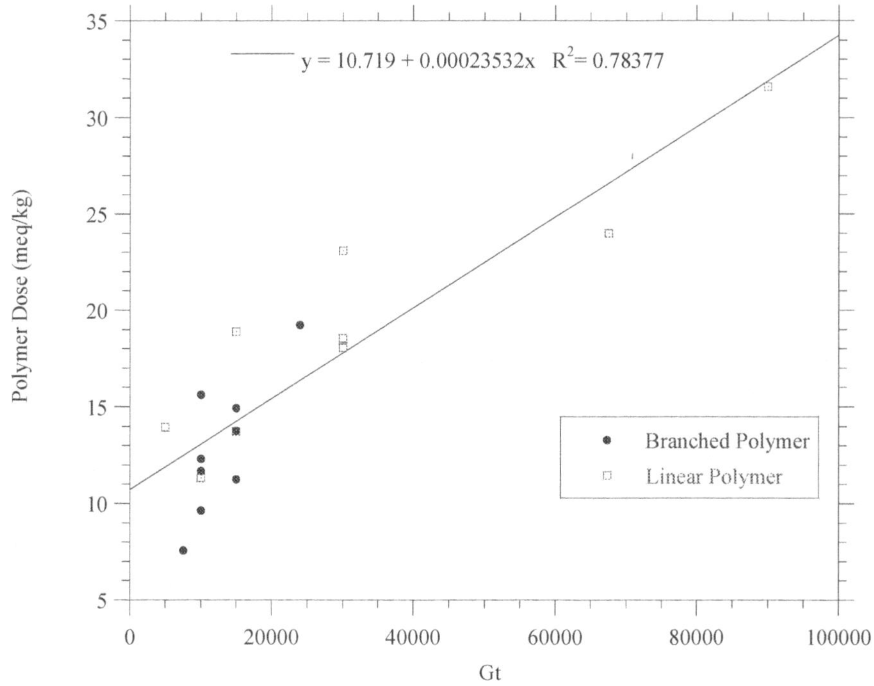

Figure 5-9. Comparison of Linear and Branched Polymer Response to Gt Values for Anaerobically Digested Sludge.

A similar study was performed to examine the effect of different G-values on a WAS; for this study, only a linear polymer was used. The results for this experiment are shown in Figure 5-10. The WAS sample was not affected by different G-values, and the polymer dose versus Gt values fit a straight line for the different G-values examined.

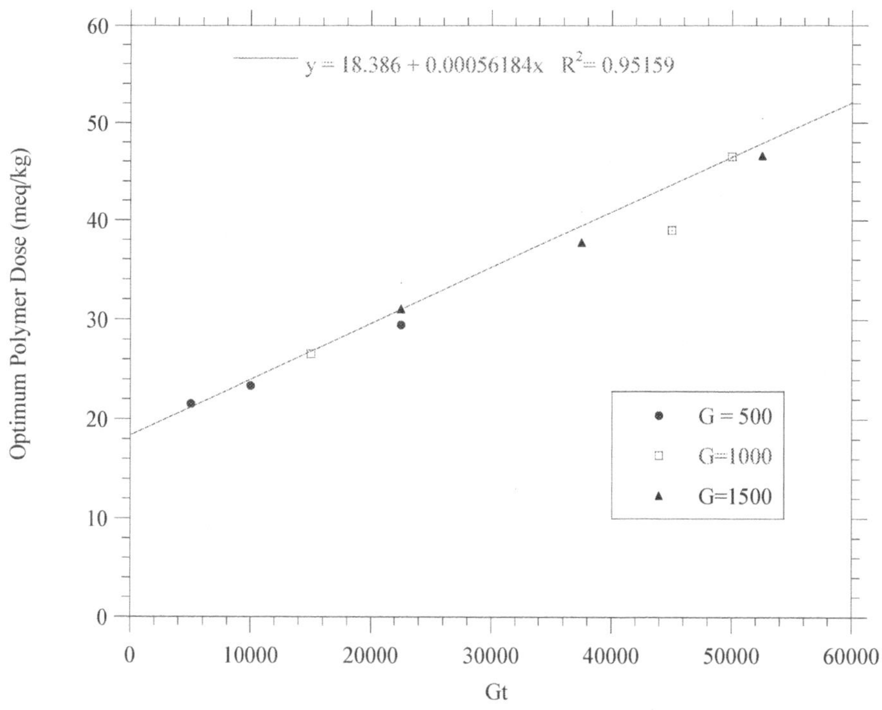

Figure 5-10. OPD vs Gt for WAS Sample Using Linear Polymer.

5.5 Evaluation of Polymer Characteristics and Their Response to Shear

A number of different polymers are available for utilities, making the choice difficult. Some of the polymer characteristics that can vary include charge, molecular weight, charge density, structure (branched or cross-linked), as well as chemistry. An important aspect of polymer selection is understanding how each polymer responds to shear. Therefore, a series of experiments were performed to examine how some of these different polymer characteristics are affected by shear. The properties examined include molecular weight, charge density, and polymer structure (linear versus structured). The results from these tests are discussed below.

5.5.1 Effect of Molecular Weight

A set of shear experiments was performed to examine the effect of MW on the response to shear during conditioning. Three polymers were used with a low (25,000), medium (400,000), and high MW (8,000,000) and an aerobically digested biosolids sample. The results from this testing are shown in Figure 5-11. Polymer MW had a significant impact on the response to shear. The low MW polymer was unable to withstand any significant shear compared to the high MW polymer. At an approximately equivalent polymer dose of 6-7 meq/kg, the low MW polymer withstood a Gt of 6000, the medium MW polymer had a Gt of 12,000, and the high MW polymer had a Gt of 165,000. Therefore, the higher MW polymer can withstand much greater shear (Gt) before the floc properties begin to deteriorate.

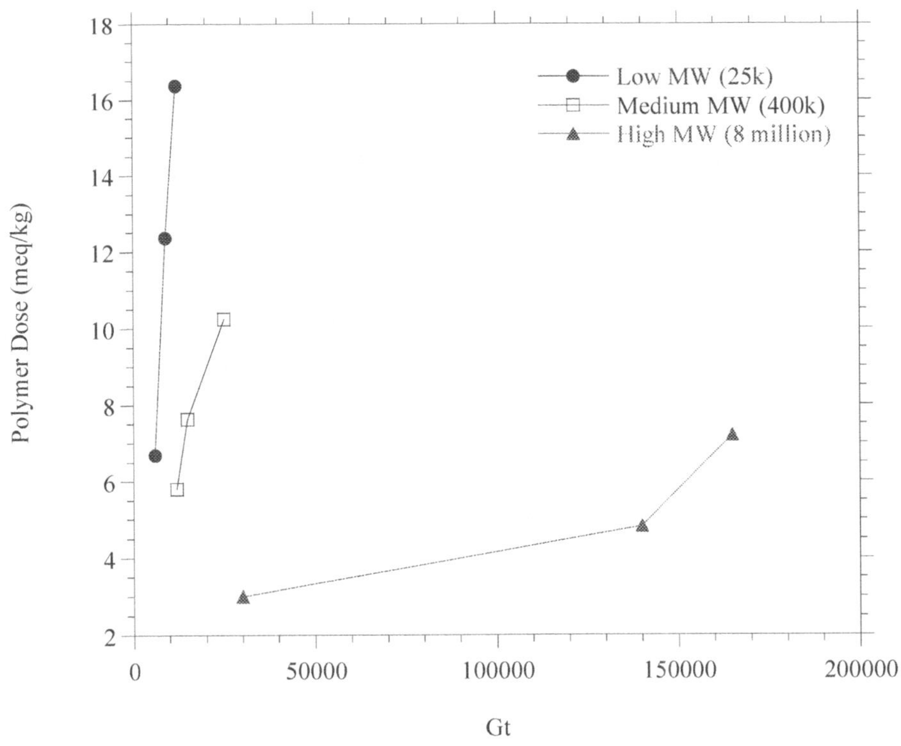

Figure 5-11. Effect of MW on Response of Polymer to Shear During Conditioning.

These results further support the role of polymer bridging in floc formation during conditioning. Polymer bridging is critical to holding flocs together once they form. For example, if charge neutralization were the sole mechanism, the addition of low MW cationic polymer would be sufficient to create flocs that are large enough and strong enough to withstand shear during dewatering. The forces holding the flocs together would include electrostatic interactions from charged sites as well as van der Waals forces. However, the inability of the low MW polymers to withstand shear (see Figure 5-11) highlight the importance of polymer bridging in floc strength.

5.5.2 Effect of Polymer Charge and Structure.

Several sets of experiments were performed to better understand how polymer charge density and structure respond to shear. For the first set of experiments, a mesophilic, anaerobically digested solids sample was obtained and dewatering experiments were performed with four different polymers. The polymer characteristics included varying structure and charge density as summarized in Table 5-2. The OPD as a function of Gt for the four different polymers is shown in Figure 5-12. The results showed that the branched polymer (2061) and linear polymer (6257) with similar MW had very similar response to shear as shown by similar slopes for the curves. The slopes of the curves for 2061 and 6257 were 0.00012 and 0.00011 g/kg, respectively. The polyacrylamide polymers, MX 60 and MX 20, had about twice the slopes, 0.00023 and 0.00025 g/kg, respectively over the same Gt range compared to the other two polymers. This was somewhat unexpected since the polymers had relatively high molecular weights and would have been expected to respond similarly to the 6257 polymer.

Table 5-2. Summary of Four Polymers Tested.

Polymer Name	Configuration	Approximate Molecular Weight	Mole Charge (%)
Cytec 2061	Branched	5 million	40
Polydyne 6257	Linear, polyacrylamide	6 million	50
Cytec MX 20	Linear, polyacrylamide	11 million	10
Cytec MX 60	Linear, polyacrylamide	11 million	40

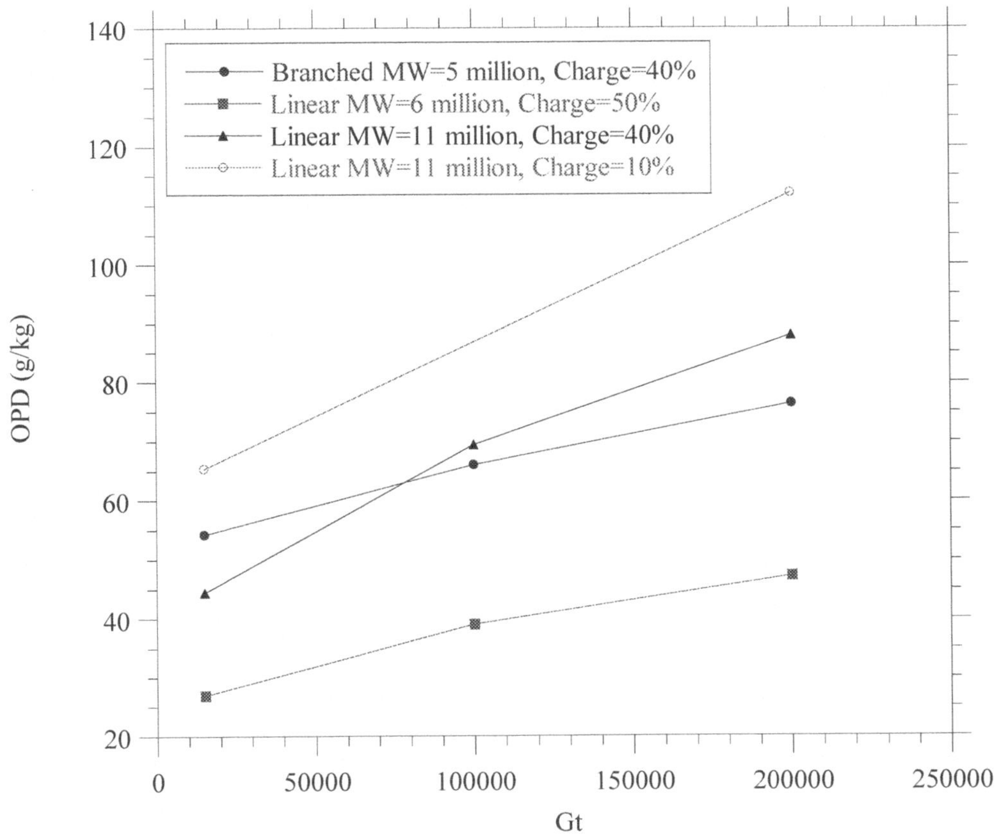

Figure 5-12. Effect of Gt on OPD for Four Different Polymers for Anaerobically Digested Sample.

A second set of tests was performed using 2061 and 6257 with a different anaerobically digested sample. Similar results were found, and the slopes of the curves were 0.00011 and 0.0001, respectively (data not shown). The results suggest that branched polymers do not have a large advantage for high shear conditioning with anaerobically digested solids. In addition, the greater charge density of MX 60 compared to MX 20 did not provide any advantage for high shear conditioning.

Additional testing was conducted to investigate the response of polymer to shear using waste activated sludges since they were shown to be one of the most shear sensitive sludges. In the first set of experiments, three polymers from Degussa Stockhausen were examined in which the polymer charge density varied and the other characteristics were similar. The OPD versus Gt curves are shown in Figure 5-13, and the OPD is expressed on a charge basis to normalize for the charge density. The results showed the different charge density polymers responded similarly to the shear, in that the slopes and magnitudes for each were very similar. The results suggest that charge density did not offer significant advantages for resisting shear.

In a similar set of experiments, polymers with similar charge densities but varying structure were examined. The polymers consisted of a linear, branched and cross-linked polymer, all from Degussa Stockhausen. The OPD versus Gt curves for these different polymer types are shown in Figure 5-14.

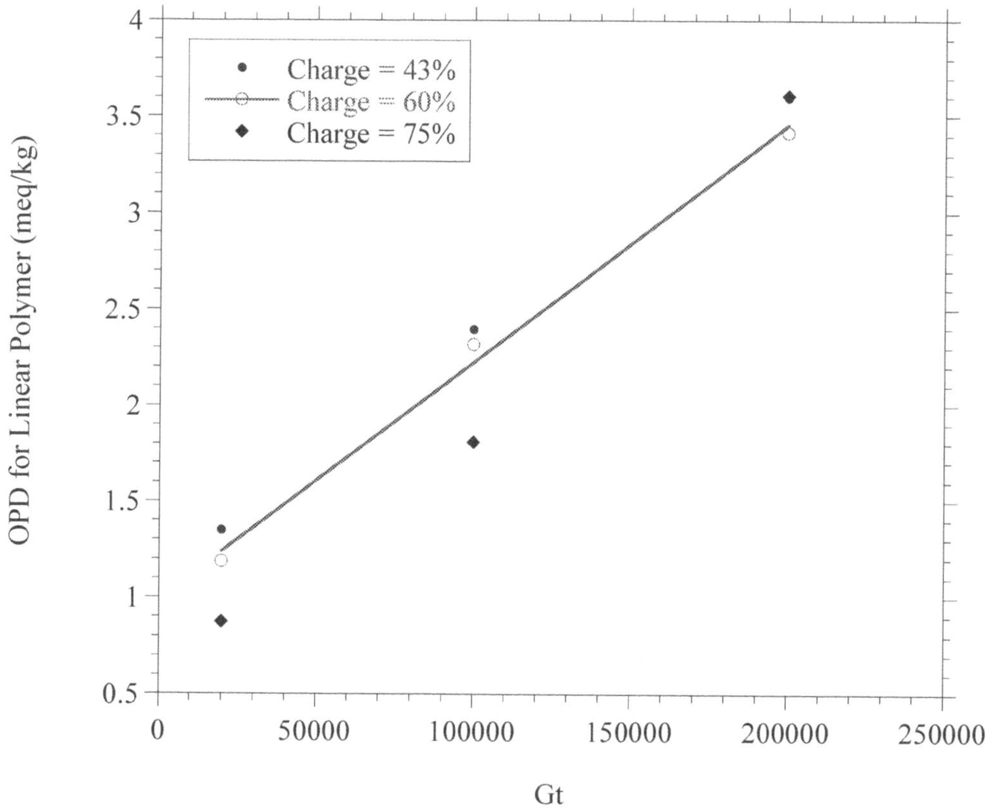

Figure 5-13. OPD versus Gt for WAS and Linear Polymers with Varying Charge Density.

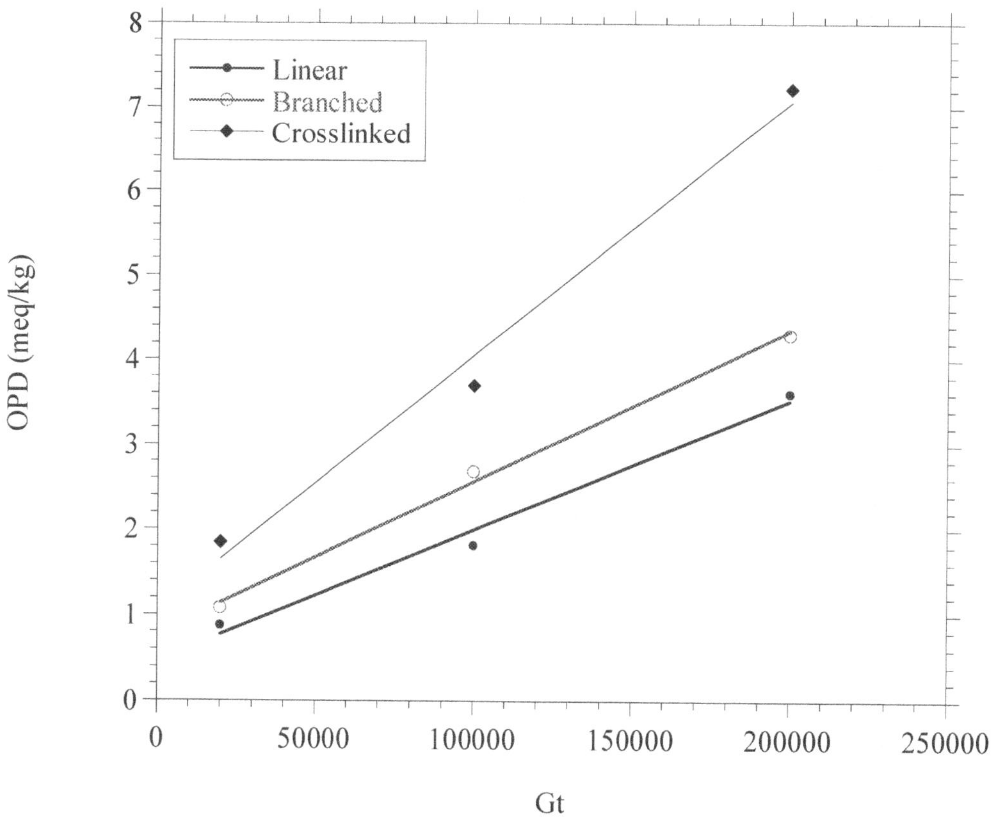

Figure 5-14. OPD versus Gt for WAS and Polymers with Varying Structure.

The results indicate that having high MW polymers is the most important factor in terms of shear resistance. The charge density and branching or cross-linking did not seem to provide consistently significant additional benefits for shear resistance in these experiments when working with high molecular weight polymers. There was some indication (Figure 5-4) that branched polymers may be able to better resist shear, and provide good performance over a range of shear values for a given polymer dose (see Figure 5-2). These tests used shear on the liquid as the measurement of shear. However, during dewatering and cake formation, the shear forces may be considerably different on a cake compared to liquid shear, and the structured polymers may perform better in terms of holding floc together during cake shear.

5.6 Other Impacts of Shear – Cake Solids and Cake Odors

The previous section clearly shows that shear applied during conditioning increases the polymer demand. Different digestion types produce solids that have different response to shear, and different polymer types also respond differently to shear. In addition, greater shear resulted in sludges that have better dewatering properties as measured by CST. However, does the increase in shear result in better dewatering in terms of cake solids? Recent research has shown that the shear also impacts odor production from the resultant cakes, can this be replicated by the laboratory procedures used in this study?

5.6.1 Effect of Shear and Polymer Type on Dewatering Properties and Cake Solids Content

To examine the dewatering properties as shear was increased, the CST measured at the OPD for the sample set using Ciba and Polydyne polymers is shown in Figure 5-15. The polymer characteristics were summarized in Table 5-2. These results are very similar to those shown in Figure 5-5. As the shear increased, the resultant CST decreased for all the polymer types examined. The lower CST values suggest the solids had better dewatering properties at the higher shear and polymer dosages. An interesting question to investigate is: does lower CST/better dewatering translate into higher cake solids content during dewatering?

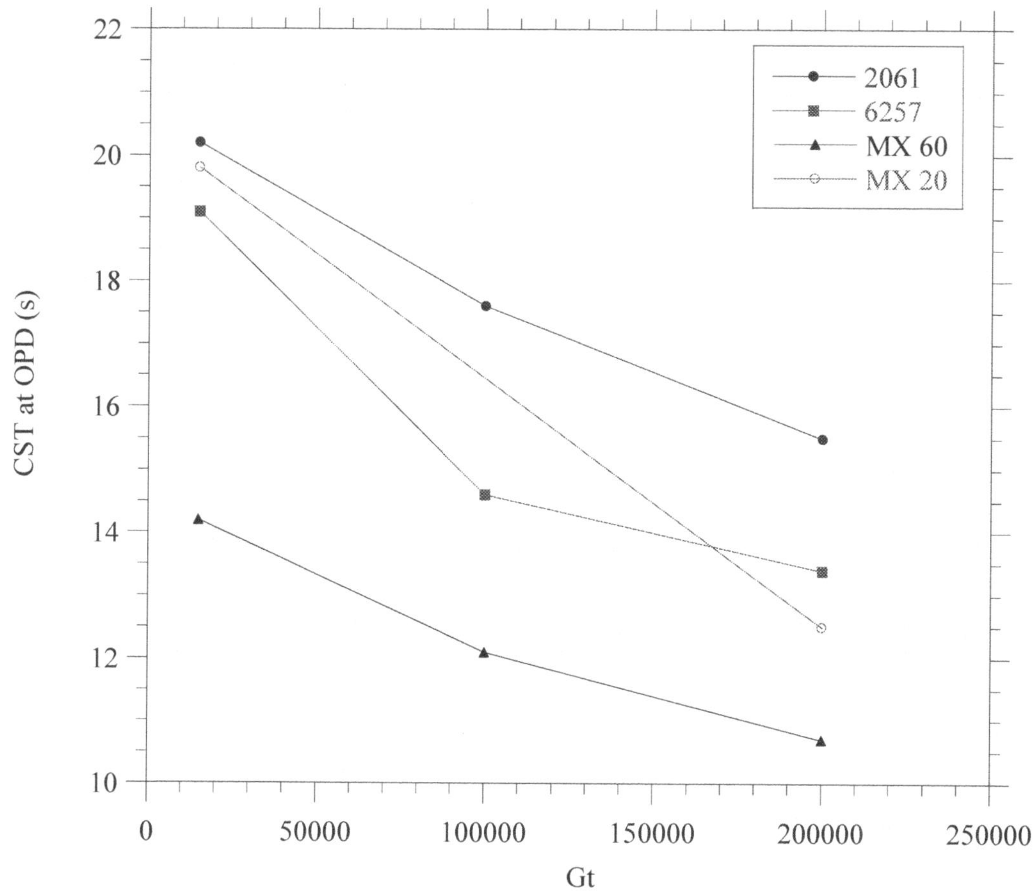

Figure 5-15. CST at the Optimum Polymer Dose for Different Gt Values and Polymer Types.

For the same set of shear experiments, the sample at the OPD was dewatered using vacuum filtration (VF) and also on a belt filter press simulator (BFP). The purpose of these experiments was to determine if shear would result in greater cake solids for the same dewatering device. The results for the VF-generated cake are shown in Figure 5-16. The increased shear resulted in an increase in the average VF cake solids concentrations for each of the polymers. For example, the 6257 polymer had an average cake solids content of 6.62% (for quadruplicate samples) for a Gt of 15,000, and this increased to 7.69% for a Gt of 100,000, and then increased further to 8.59% at the highest Gt value of 200,000. These increases were statistically significant ($p < 0.05$) when compared to the Gt of 15,000, although the cake solid values at 100,000 and 200,000 Gt were not significantly different for both the 2061 and MX 60 polymers. All the cake solids were significantly different for the 6257 and MX 20 polymers. The MX 20 polymer had the greatest increase in cake solids for the two samples. This polymer had the lowest charge density of the polymers tested, but it is not clear why this would impact cake solids to a greater extent.

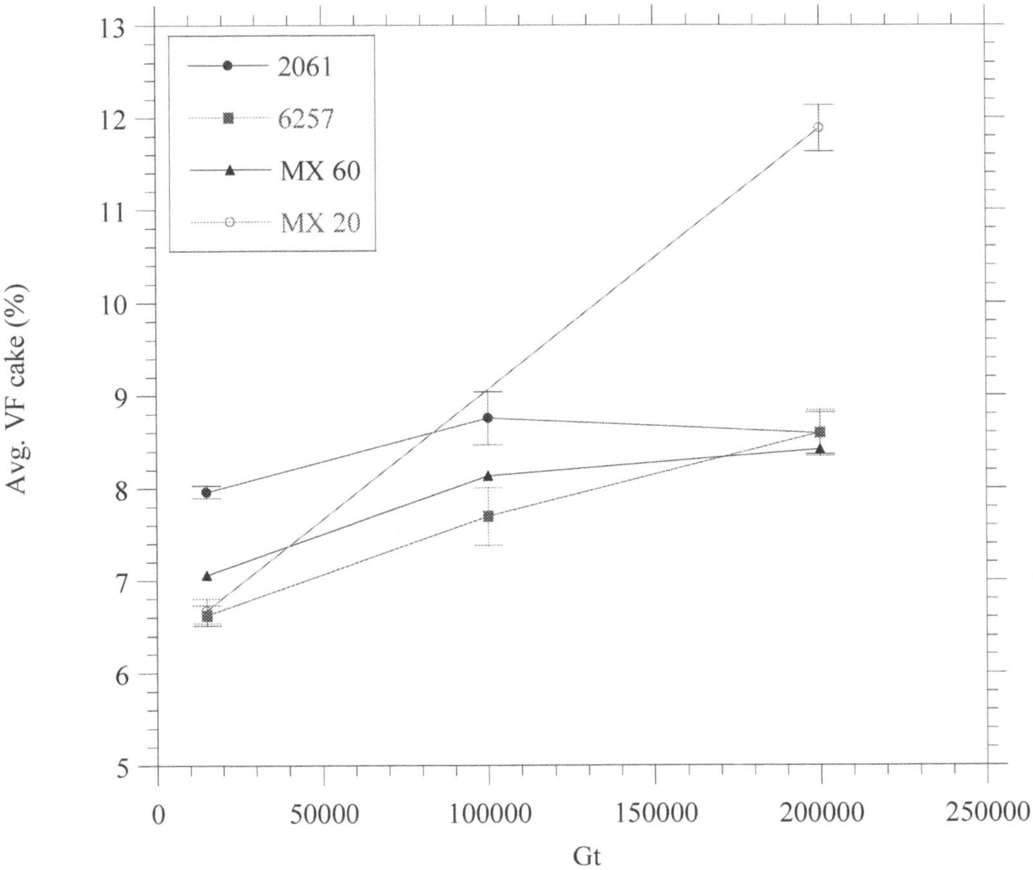

Figure 5-16. Effect of Gt on the Average Cake Solids Content from Vacuum Filtration (Error Bars Represent one Standard Deviation. For Clarity, not all Error Bars are Shown).

The increase in cake solids was greater for the belt filter press simulator, as shown in Figure 5-17. As shear (and the associated polymer dose) increased, the average cake solids content also increased for all the different polymers. The rate of increase was similar for the different polymers. The results suggest that increased shear during conditioning (prior to dewatering) may be a strategy to increase cake solids and thereby reduce disposal costs. This would have to be balanced with the greater polymer costs, but results show that this may be a method to achieve greater cake solids.

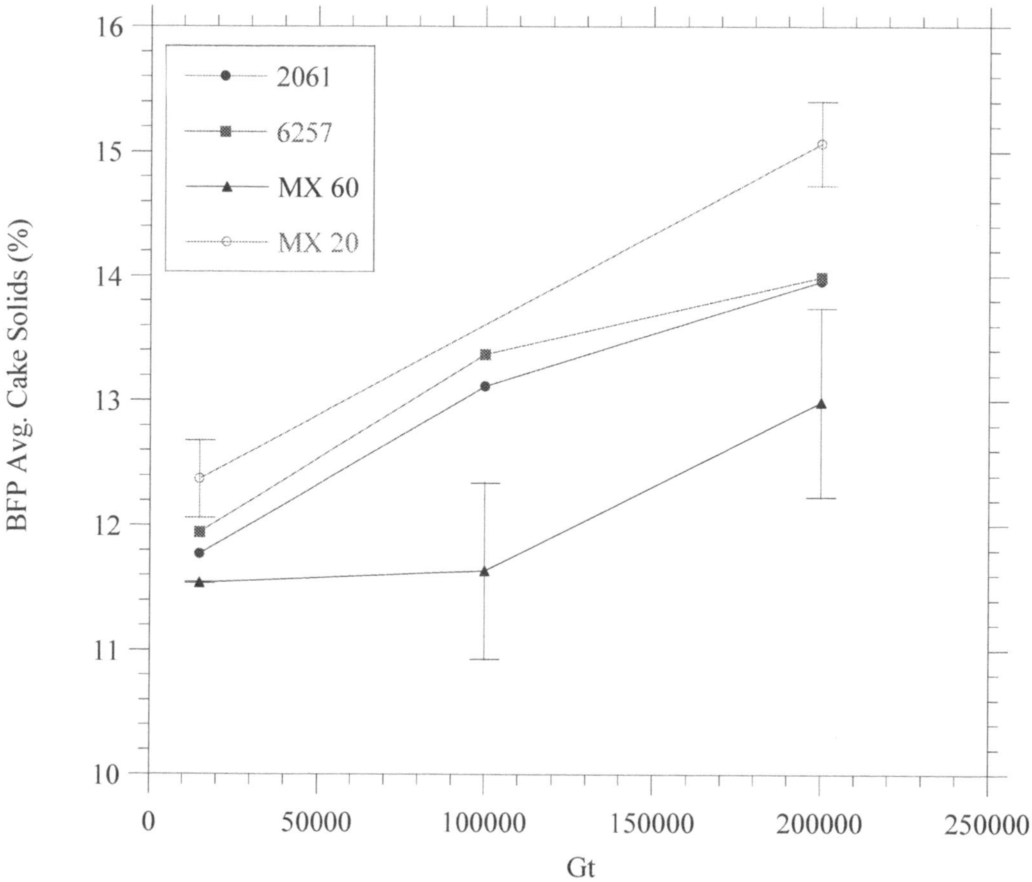

Figure 5-17. Effect of Gt on the Average Cake Solids Content from a Belt Filter Press Simulator, Note Error Bars Represent one Standard Deviation, for Clarity, not all Error Bars are Shown.

For all the polymers except MX 60, a good linear relationship was found between CST and cake solids at the OPD (Figure 5-18); as the CST at the OPD decreased, the cake solids from the BFP increased.

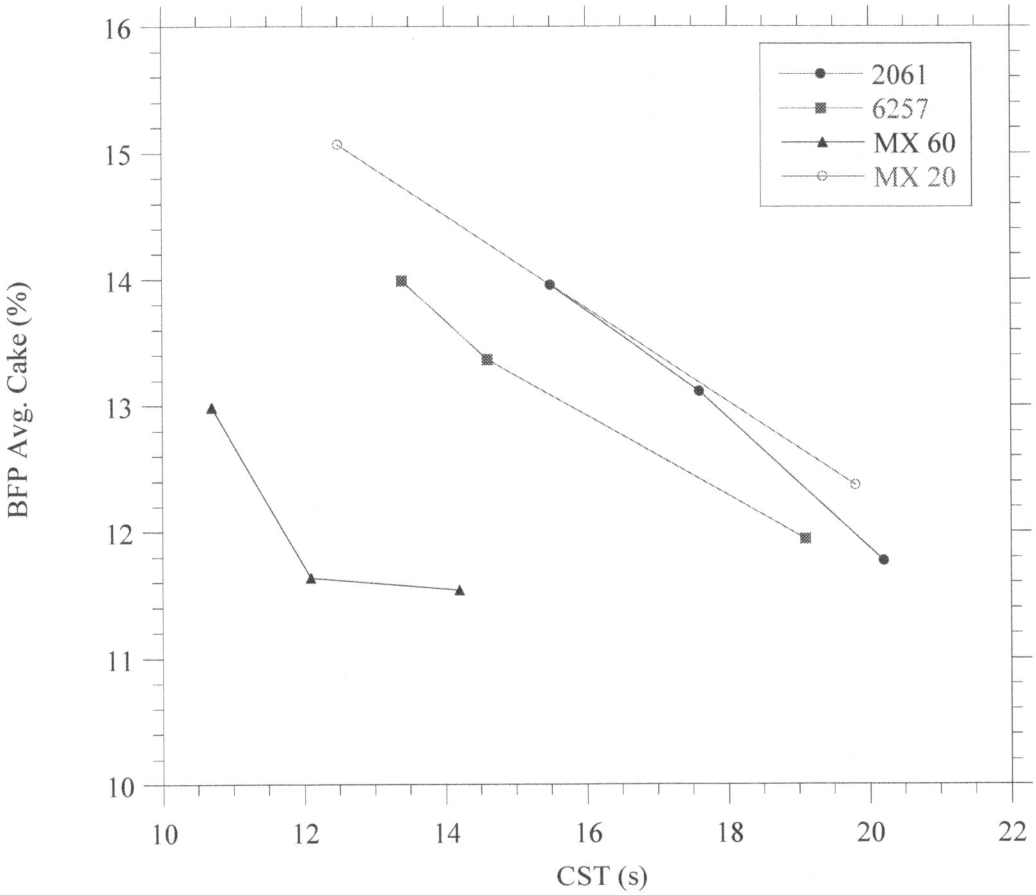

Figure 5-18. Relationship Between BFP Cake Solids and CST at the OPD for Different Polymers.

The increase in cake solids appears to be related to a break-up and reformation of the floc that results in release of floc-entrapped water. This water could be partly bound water displaced by the polymer as well as water entrapped in the pores of the floc. Microscopic analysis of the flocs produced using different shear showed that flocs became more compact or smaller in size as the shear was increased (e.g., Fig. 5-19). The unconditioned sample had small, low-density/transparent, amorphous flocs. Conditioning resulted in larger, more densely packed flocs. As the shear increased, the overall floc size became smaller. Although the pictures do not clearly show the smaller sized particles, the observations of the flocs throughout the microscopic field showed this trend clearly.

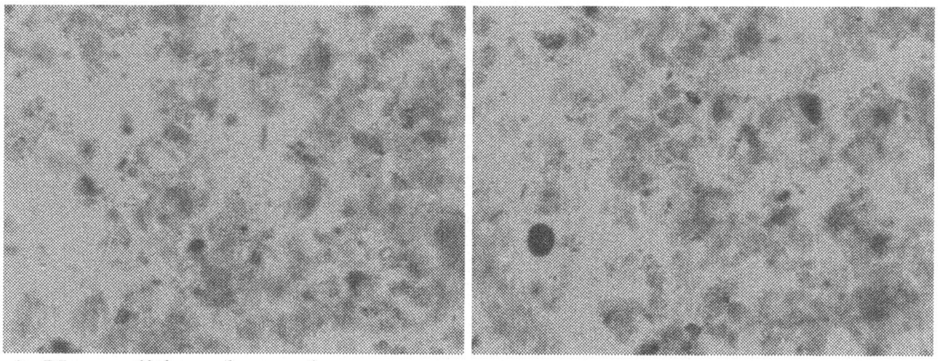

a) Unconditioned sample

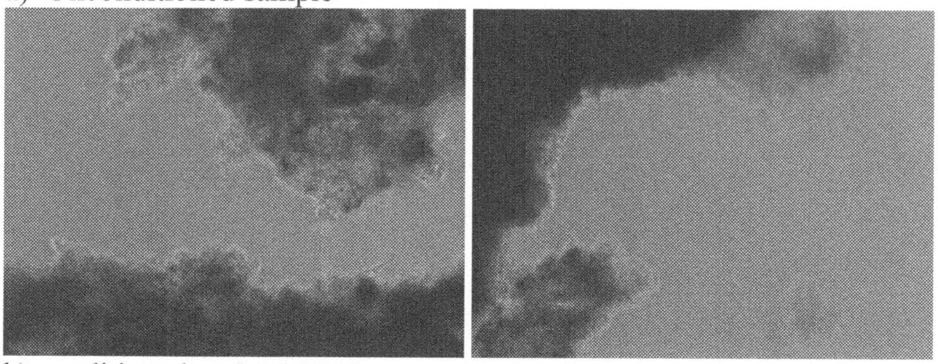

b) conditioned at Gt=15,000

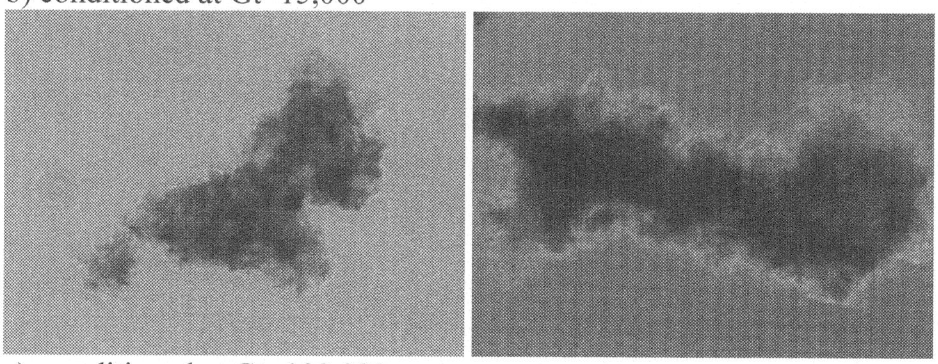

c) conditioned at Gt=100,000

Figure 5-19. Micrographs of Solids Samples Conditioned Using Different Shear (Magnification 100x for all Pictures).

5.6.2 Impact of Shear on Odor Production from Biosolids Cake

Recent research has shown that shear and polymer dosage may be associated with cake odor formation (Higgins et al., 2005a). Several studies were performed to examine the effects of shear and of polymer type on odor production as part of this project. Shear experiments were performed in which Gt values of 60,000 and 100,000 were applied during conditioning. After this shear, the solids were dewatered in a laboratory centrifuge, and the cake samples were stored in serum bottles. Each day, headspace samples were collected and analyzed for odor-causing compounds, mainly methyl mercaptan (MT) and dimethyl sulfide (DMS)(see Higgins et al., 2005b for more detailed methodology). The effect of these two different conditioning shear values is shown in Figure 5-20. Interestingly, no detectable MT was recorded for these experiments. In addition, cakes produced from the same anaerobically digested biosolids that were dewatered on a full-scale high solids centrifuge had peak concentrations of MT near 500 ppmv. A second experiment was performed in which the same shear/mixing regime was applied, and then the cake generated after centrifugation was further sheared by a meat grinding apparatus. This system had a scroll conveyor which pushed the solids forward, and then extruded them through small openings at the end of the scroll. It was believed that this cake shear would be similar to the shear in a high solids centrifuge. This sheared cake was then placed in a serum bottle and analyzed as before. The results are also shown in Figure 5-20. This cake shear regime produced an odor generation pattern similar to the full-scale sample, although the actual concentrations were higher for the laboratory sample. The shear during conditioning, or liquid shear, did not cause production of odorous compounds, but rather the shear during dewatering or shear applied to the cake seemed more important.

This analysis shows the importance of distinguishing between shear during conditioning and shear during dewatering or cake shear, which seem to have very different impacts on the cake properties. The cake shear method used in this experiment produces a cake with similar odorant profiles as full-scale high solids centrifuges. This method can now be applied to study factors which may affect odorant production, such as polymer dose, shear during conditioning, and polymer type. This research is very important to link polymer demand and characteristics as well as shear to the product quality in terms of solids and odors.

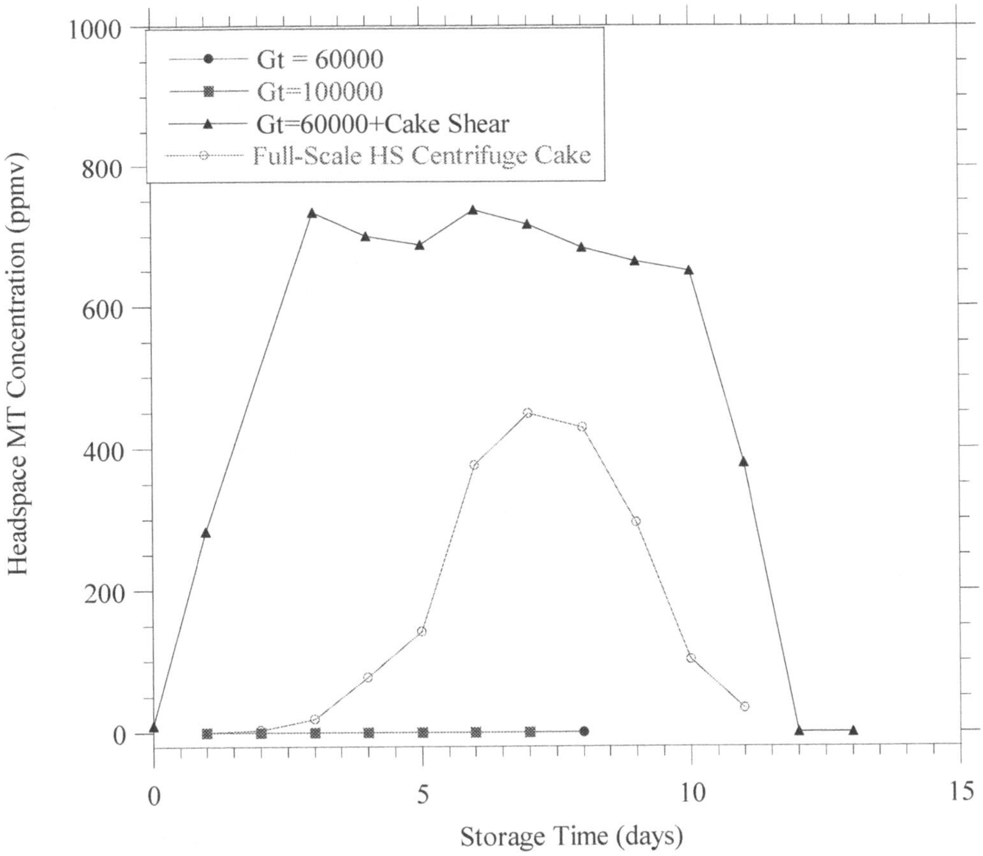

Figure 5-20. Effect of Shear on Odor Production from the Cake.

Using this approach, the effect of different polymers on odor production from the cake was studied. Research has shown that polymer binds protein and deposits this protein in the cake. Cake protein then can be degraded to form odorous compounds. Therefore, polymers with different abilities to bind protein may produce cakes with different odor profiles. Results from one trial comparing the branched 2061 polymer with the linear 6257 polymer are shown in Figure 5-21. The two polymers produced cakes with very similar odorant profiles, suggesting that odor production is not a deciding factor when choosing between these polymers. Additional research is needed to better understand how polymers affect odor production, especially in relation to their response to shear during conditioning, and shear of the cake.

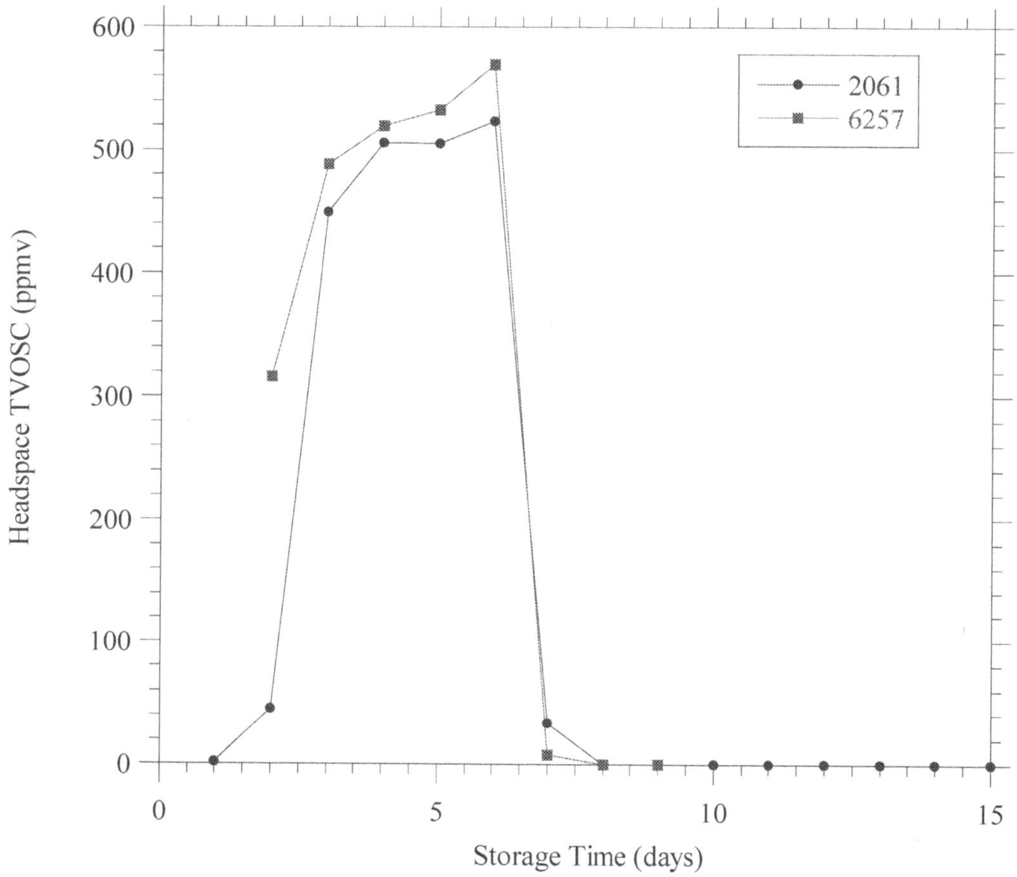

Figure 5-21. Comparison of Total Volatile Organic Sulfur Compounds (TVOSC) Produced from Cakes Produced by Conditioning with Two Different Polymer Types.

5.7 Conclusions and Implications

5.7.1 Conclusions

- Shear has a significant impact on polymer demand, and different polymer and sludge types respond differently to shear.
- Greater amounts of shear increase the polymer demand for conditioning, but the effect is different for different types of sludges. Aerobically and anaerobically digested biosolids behave similarly, and are least impacted by shear, followed by TPAD and thermophilic biosolids. WAS had the greatest increase in OPD as a function of Gt but was similar to the thermophilic biosolids. The results are based on limited testing for a few sludge types. Additional testing is needed to determine if these relationships hold true for additional sludge types.
- For digested sludges, the rate of increase in OPD as a function of Gt (the slope of the OPD-Gt curves) was related to the initial biocolloid concentration of the suspension. The results showed thermophilic digestion released greater amounts of biocolloids and the flocs from thermophilic treatment are more susceptible to breakup by shear.
- The greater the MW of the polymers, the better they were able to withstand the effects of shear. These results supported the role of polymer bridging in forming strong flocs that do not break up during dewatering.
- For higher MW polymers (greater than about one million), variations in structure and charge density did not appear to have large impacts on resistance to shear. Some of the results suggest that branched polymers may provide a benefit for shear resistance over linear polymers, but the results were not conclusive. The key factor in resistance to shear was having high MW polymers.
- The use of Gt as a parameter to simulate shear during the conditioning and dewatering process may not fully replicate actual shear during dewatering. The results from this work suggest different effects between shear applied to the liquid and shear applied to the cake. Currently, the Gt concept which measures or imparts shear to the liquid is the most accepted tool for replicating shear during dewatering; however, additional research is needed to better understand cake shear and its role in impacting polymer demand and cake quality parameters such as odor characteristics.

5.7.2 Implications

- The results from this study and others can help utilities better predict the impact of digestion on OPD. For example, a plant considering changing from mesophilic to thermophilic anaerobic digestion will likely experience a significant increase in polymer demand due to greater amounts of biocolloidal matter in solution. In addition, the digested product will be more susceptible to shear. Therefore, if the plant is using high shear dewatering (such as a high solids centrifuge), the resulting polymer demand will be greater compared to a belt filter press.
- The impact of shear also can be used to estimate the change in polymer demand associated with different dewatering devices. For example, if the representative Gt of

different dewatering devices is known, OPD vs Gt curves for the particular sludge can be generated to predict the polymer demand under different shear conditions. In addition, the effect of different polymer types can be investigated by developing a family of Gt curves for the different polymers.

- The variability in shear response of different sludge and polymer types creates additional complexity in developing a universal model to predict polymer demand, which is discussed in more detail in Chapter 7.0.

CHAPTER 6.0

GT OF FULL-SCALE DEWATERING EQUIPMENT

6.1 Background

The Gt value of full-scale dewatering equipment can be estimated by correlating the polymer dose-response profile of full-scale dewatering equipment with the Gt value of a calibrated laboratory mixer using the same polymer and solids to obtain similar dewatering properties (Murthy et al., 1997; Novak and Lynch, 1990). In these experiments, the same polymer dose used in the full-scale device is used in the laboratory mixer at a defined G. The system is mixed for varying Gt values to obtain similar dewatering characteristics as in the full-scale trial. This information is important in understanding the polymer demand exerted by different dewatering devices as a result of the shear that is imparted during dewatering. Research by Novak and coworkers has determined the Gt values of several different full-scale dewatering devices. A summary of these results are provided in Table 6-1. Novak and co-workers surmised that most of the shear in these devices occurred as a result of the movement of water through the developing cake, and removal of water prior to pressure application would reduce the shear (such as in the free drainage zone of a belt filter press).

Table 6-1. Summary of Gt Values for Different Full-Scale Dewatering Devices.

Dewatering Device	Gt	Reference
Fixed Volume, Recessed Chamber Filter Press	30,000	Novak and Lynch (1990)
Belt Filter Press	10,000	Novak et al. (1993)
Screw Press	10,000	Murthy et al. (1997)
Centrifuge - Low Solids	10,000	Murthy et al. (1997)

6.2 Gt of Full-Scale Dewatering Equipment

Advances in centrifuge technology have led to the development of high-solids centrifuges which are becoming common at larger municipalities. These centrifuges can produce cakes greater than 30% from anaerobically digested biosolids. High-solids centrifuges are generally thought to impart greater amounts of shear than other centrifuges or presses, however, no literature is available on the typical Gt values of high and medium solids centrifuges.

Trials were performed to determine the approximate Gt values for several different dewatering devices. These trials used the approach of Novak and coworkers in which the OPD was determined for the dewatering device and then this dose was used to determine the corresponding Gt value using a standardized mixing setup with the same polymer and feed as the field trial. The trials used two methods for determining the OPD. First, field trials were performed in which the investigators visited the site and determined the polymer dose profile through testing of different dosages. From this, the OPD was determined using both cake and capture data. Secondly, the OPD was reported by the utility and no field testing by the investigators was performed. All samples were anaerobically digested biosolids. (Despite the focus on OPD, in practice there typically is a range for polymer dosage that can provide desired dewaterability.)

A summary of the results are provided in Table 6-2. The two high solids centrifuges had the greatest shear, with Gt values between 75,000 and 120,000. The medium solids centrifuge had a moderately high shear and the belt filter press had the lowest shear. The belt filter press value compared well with previous research (see Table 6-1). The OPD, on an equivalent basis, is also provided in the table. The higher shear devices generally had greater polymer requirements, although the belt filter press had a relatively high OPD despite the low Gt. This is likely a result of the relatively high biocolloid concentration measured in this sample (1500 mg/L) compared to the other samples, which were all below 1000 mg/L. This results shows that multiple factors can affect polymer demand of sludges, which adds to the complexity of the analysis.

Table 6-2. Comparison of Gt Values for Different Dewatering Devices Analyzed in This Study.

Dewatering Device	Feed Sludge Type	Estimated Gt	OPD (meq/kg)	OPD Data Source[1]
Sharples High Solids Centrifuge	Anaerobically Digested	120,000	59.3	Self-reported
Humbolt High Solids Centrifuge	Anaerobically Digested	70,000	29.1	Field Trial
Bird Medium Solids Centrifuge	Anaerobically Digested	30,000	19.6	Field Trial
Belt Filter Press	Anaerobically Digested	10,000	42.3	Self-reported

[1]Self-reported refers to OPD data provided by the utility, field trial refers to OPD determined by field testing by the utility and principal investigators.

A compilation of Gt values determined in this study as well as other studies is shown in Figure 6-1. Since the polymer dose is directly related to the Gt or shear, this graph gives an indication of the expected changes in polymer dose for different types of dewatering equipment. This information can be used to help better understand the implications of changing from one dewatering device to another. For example, changing from a belt filter press to a high solids centrifuge could increase polymer demand significantly.

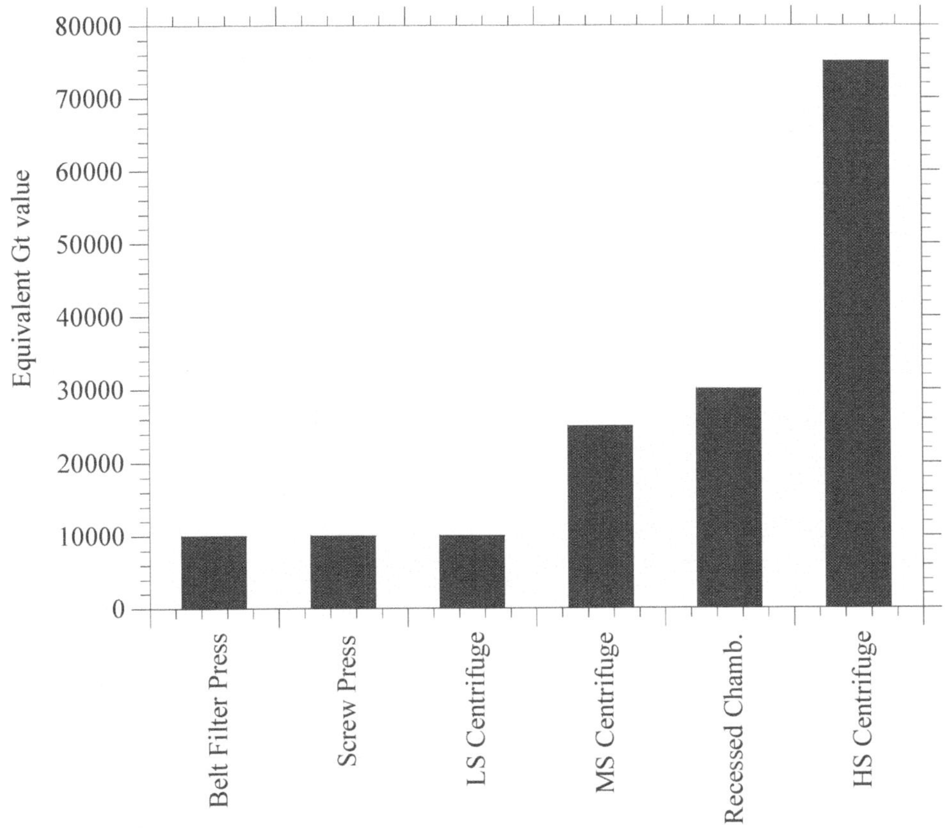

Figure 6-1. Comparison of Gt Values for Dewatering Equipment.

6.3 Conclusions and Implications

6.3.1 Conclusions

The shear associated with full-scale dewatering devices can be estimated with the equivalent Gt by using techniques previously described in the literature. Using these techniques the shear associated with medium and high solids centrifuges were estimated. The medium solids centrifuge had an estimated Gt of 30,000, while the two high-solids centrifuges had Gt values between 70,000 and 120,000. This data is useful in predicting polymer demand due to different dewatering equipment.

6.3.2 Implications

Gt values can be used to estimate the polymer demand associated with different types of dewatering equipment. Many large plants are currently upgrading to high solids centrifuges and this change in dewatering equipment could have a significant impact on the polymer demand and subsequent costs. Considering Figure 6-1, changing from a belt filter press to a high solids centrifuge could more than double polymer demand; assuming similar relative costs, the polymer costs would increase by the same amount.

CHAPTER 7.0

RESEARCH APPLICATIONS: CAN WE PREDICT POLYMER DEMAND?

7.1 Development of a Universal Predictive Model

The research to date has confirmed that polymer demand for sludge conditioning is a function of at least three factors:

- Biocolloid concentration;
- Polymer characteristics, especially polymer charge density (PCD); and
- Shear.

Based on this understanding, these factors could be combined to potentially develop an empirical model that could be used to estimate polymer demand. However, the results from the research also showed that different sludges and polymer types respond differently to shear, which can add complexity to the model. Therefore, several simplifying assumptions were made in the model. First, it was assumed that the model would only consider high MW polymers since the lower MW polymers were not practical for conditioning. In addition, the final model only considered a simple Gt term for all cases, which is explained in more detail below. The model was developed with the following data set:

- Sludges of differing biocolloid concentration (n=21);
- Polymer types with different polymer charge densities (PCDs), and structure (n=20);
- Shear or Gt values which ranged from 5,000 to 200,000 (n=22).

The data set provides 66 data points that can be used in the model, although some models had slightly more or less data depending on which parameters were measured for each sample type. A number of multi-parameter regression analyses were performed using the data and the different parameters. A relatively good fit to the data was achieved using just the parameters of Gt, biocolloid concentrations, and polymer charge density. Possible measures of biocolloid concentration include SCOD or protein plus polysaccharide, and four different particle sizes. All of these biocolloid measures were examined in analyzing the data for an appropriate model.

Based on the regressions for the different models examined, the simplest model or equation for predicting polymer demand, with the best correlation coefficient was:

$$\text{Polymer Dose} = -7.61 + 0.00692(\text{SCOD}) + 16.0(\text{PCD}) + 3.99\times10^{-4}(\text{Gt}) \qquad \text{(Equation 7.1)}$$

where:

SCOD = concentration of soluble chemical oxygen demanding constituents less than 4.2 µm (mg/L);

PCD = polymer charge density (meq/g);

Gt = unitless mixing intensity parameter;

Polymer Dose = meq/kg DS.

The correlation coefficient (r^2) for this equation was 0.73, indicating a reasonably good fit. This model also uses the simplest parameter for measuring biocolloid concentration, which was SCOD concentration less than 4.2 µm. The correlation coefficient can be thought of as the proportion of variability in the polymer dose estimate that is explained by the equation. Therefore, this model accounts for 73% of the variability in the data. Additional analysis showed that all three parameters are significant to the model.

One of the difficulties for developing this model is incorporating the effect of shear. The simple approach of using the Gt value assumes that different polymers and different sludge types respond the same to shear, but the results of this research have shown this is not the case. For example, the results discussed in Chapter 5.0 showed that the slope of the Gt-OPD curves varied from about $2.0x10^{-4}$ for anaerobically and aerobically digested sludges to about $5.6x10^{-4}$ for the thermophilic and WAS samples. The model includes an average value for the Gt term of $3.99x10^{-4}$; using this value for all solids types may lead to some errors in its implementation. Despite this shortcoming, the model does a reasonable job of fitting the data. A number of other model runs were performed with different parameters, such as assigning a Gt-OPD specific slope to the different sample types. These additional parameters increased the complexity of the model without appreciably increasing the correlation coefficient.

Additional analyses were performed using the four biocolloid size ranges independently to determine if the different sizes may produce models with better fit to the data. The protein plus polysaccharide concentration was used as the measure of biocolloid concentration. From the analysis, the correlations were not as good as the previously explained equation using SCOD, with r^2 values that ranged from 0.60 - 0.70 depending on the size fraction used (Table 7-1). A similar regression analysis was performed using SCOD as the measure of biocolloid concentration, and these models produced similar results, with r^2 values between 0.64 - 0.65.

Table 7-1. Regression Results for Multi-Parameter Model to Predict Polymer Demand.

Biocolloids Size Fraction (µm)	Parameter coefficients				Correlation Coefficient, r^2
	Protein+Polysaccharide (mg/L)	PCD (meq/g)	Gt	Constant	
4.2	0.0095	20.4	3.96x10-4	-7.08	0.70
8.0	0.0142	16.9	4.64x10-4	-10.6	0.60
16.0	0.0148	16.6	4.46x10-4	-10.4	0.64
22.5	0.0157	17.4	4.42x10-4	-11.8	0.63

These results show that a predictive model can be developed to estimate the polymer demand for a sludge given the biocolloid concentration, polymer charge density, and Gt value of the dewatering device. The correlation coefficient is reasonably good, but more research is needed to fully develop and refine the model. The model also shows how the different factors contribute to polymer demand and methods to reduce the polymer demand can focus on shear, biocolloids, and polymer characteristics.

7.2 Comparison of Universal Model with Field Data

A full-scale field trial was performed at the Philadelphia Water Department (PWD) to investigate how well the model predicted polymer demand under varying conditions. A mannich type polymer and a non-mannich formulated polymer were used to dewater two different anaerobically digested solids on both a high solids centrifuge and medium solids centrifuge. The two polymers were Cytec polymers with the following trade names:

1) WP 2004 (Abbreviation 2004)

2) SUPERFLOC C-1596 (Abbreviation 1596)

The WP 2004 polymer is described by Cytec as a methyl chloride quaternary ammonium salt or a "Mannich Quat" and the 1596 polymer is described as a resin amine polymer. The polymers had charge densities of 0.831 and 0.624 meq/g for the 2004 and 1596 polymers, respectively.

The full-scale tests included eight different trials (two polymers, two sludges, and two centrifuges) and for each trial, 3-5 different polymer doses were examined. As a result, a total of 27 different tests were performed. In each test, the polymer/sludge/centrifuge combination was initiated, and once the system had stabilized, cake and centrate samples were collected. The centrate samples were analyzed for total solids as well total COD, total protein, SCOD, and soluble protein. The cake samples were analyzed for solids content. Using this data, the OPD was determined. In several cases, the OPD was not clear from the data, and it may have been slightly higher than the dosage range used, although in all cases, the solids capture was good and cake solids were in a desirable range. An example set of data for a trial using the HS centrifuge is shown in Figure 7-1.

The data also allow for a direct comparison of the two polymers. For example, the centrate TS for the high solids centrifuge and medium solids centrifuge as a function of polymer dose for the two polymers is shown in Figures 7-2 and 7-3, respectively. The polymer dose for each is expressed on a charge basis for direct comparison. For the high solids centrifuge, both polymers provided similar centrate quality at equivalent dosages, Figure 7-2. However, for the medium solids centrifuge, the capture was better using the 1596 polymer compared to the 2004 polymer (Figure 7-3).

To compare the field data with the model predictions, the model equation was used to calculate the OPD given the appropriate inputs of biopolymer, Gt, and polymer charge. A summary of the input parameter values and calculated OPD from the model is provided in Table 7-2 as well as the OPD determined from the field data. As shown in the table, the model generally did a reasonable job of predicting the OPD for the medium solids centrifuge, but overpredicted the OPD for the high solids centrifuge. The percent differences in the actual versus predicted OPD values ranged from 0-14% for the medium solids centrifuge, and 33-85% for the high solids centrifuge.

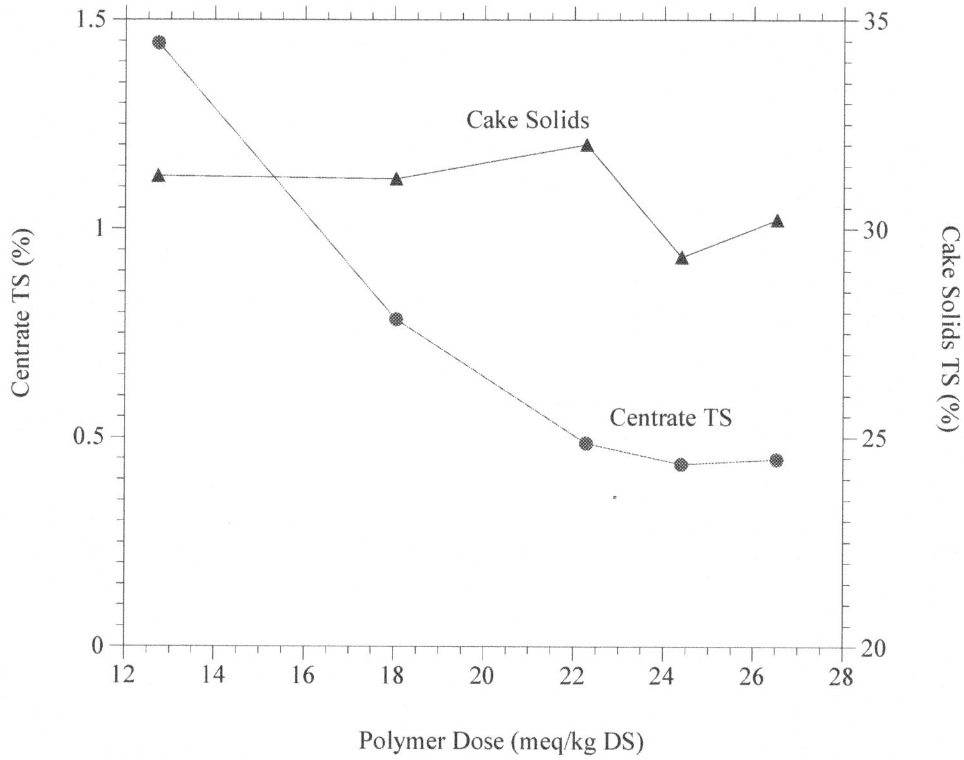

Figure 7-1. Centrate TS and Cake Solids as a Function of Polymer Dose for an Experimental Full-Scale Trial.

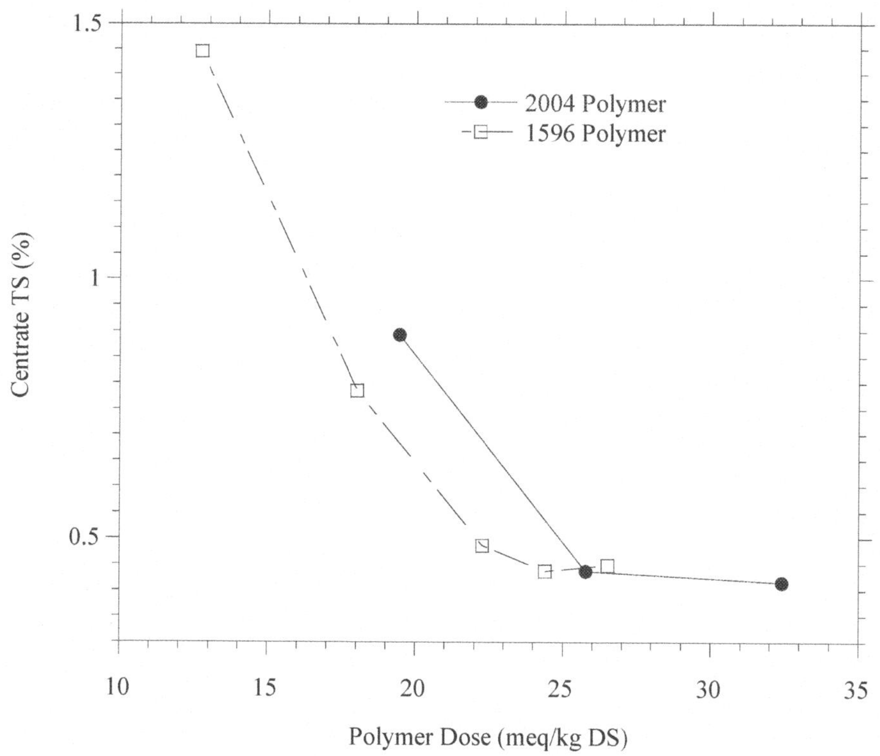

Figure 7-2. Comparison of Polymer Performance in Terms of Centrate TS for the High Solids Centrifuge and NE Sample.

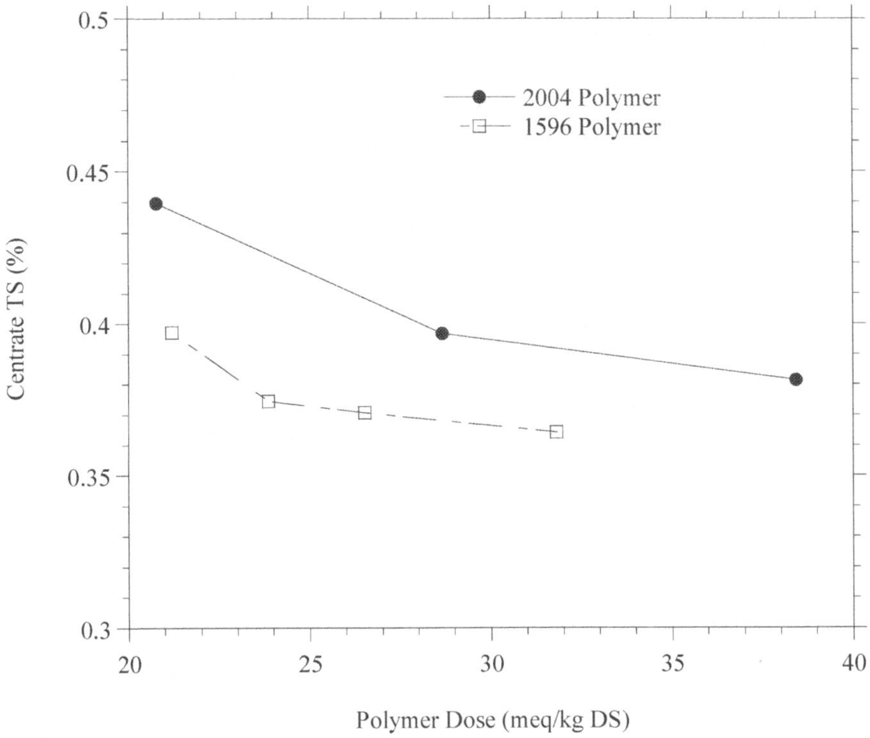

Figure 7-3. Comparison of Polymer Performance in Terms of Centrate TS for the Medium Solids Centrifuge and NE Sample.

Table 7-2. Summary of Input Values for Model, and Calculated OPD for the Given Conditions.

Sample	Polymer	Dewatering Device	PCD (meq/g)	SCOD < 4.2 mm (mg/L)	Gt	Model OPD (meq/kg)	Actual OPD (meq/kg)
PWD NE	Cytec 2004	HS Centrifuge	0.831	1356	70000	43.0	32.4
PWD NE	Cytec 1596	HS Centrifuge	0.624	1356	70000	39.7	26.5
PWD NE	Cytec 2004	MS Centrifuge	0.831	1356	30000	27.1	28.7
PWD NE	Cytec 1596	MS Centrifuge	0.624	1356	30000	23.7	26.5
PWD SW	Cytec 2004	HS Centrifuge	0.831	928	70000	40.0	21.6
PWD SW	Cytec 1596	HS Centrifuge	0.624	928	70000	36.7	21.2
PWD SW	Cytec 2004	MS Centrifuge	0.831	928	30000	24.1	21.2
PWD SW	Cytec 1596	MS Centrifuge	0.624	928	30000	20.8	20.8

The results show the model has some promise for predicting the OPD; however, additional research may be needed to fine-tune the model to more accurately predict polymer demand. In addition, the results may be better applied on a site-specific basis. For example, the Gt-OPD curve could be developed at a given plant, and this would eliminate the variability in the model associated with different sludge types and their response to shear. Although a site-specific approach would require additional testing, such testing could be completed in a day and would provide valuable insight into how a plant's solids respond to shear.

7.3 Model Applications

The predictive equation can be used to evaluate process changes, including use of different dewatering equipment, that might affect polymer demand. The model equation was used to estimate the polymer dose for an anaerobically digested biosolids and the different dewatering equipment for which equivalent Gt values have been determined (Figure 7-4). From this analysis, plant personnel can evaluate potential impacts of upgrading from a belt filter press to a high solids centrifuge. The polymer demand could increase by a factor of two or more. Although the model is still new and will need further development and refinement, this type of model does provide the basis for examining polymer demand impacts under different situations.

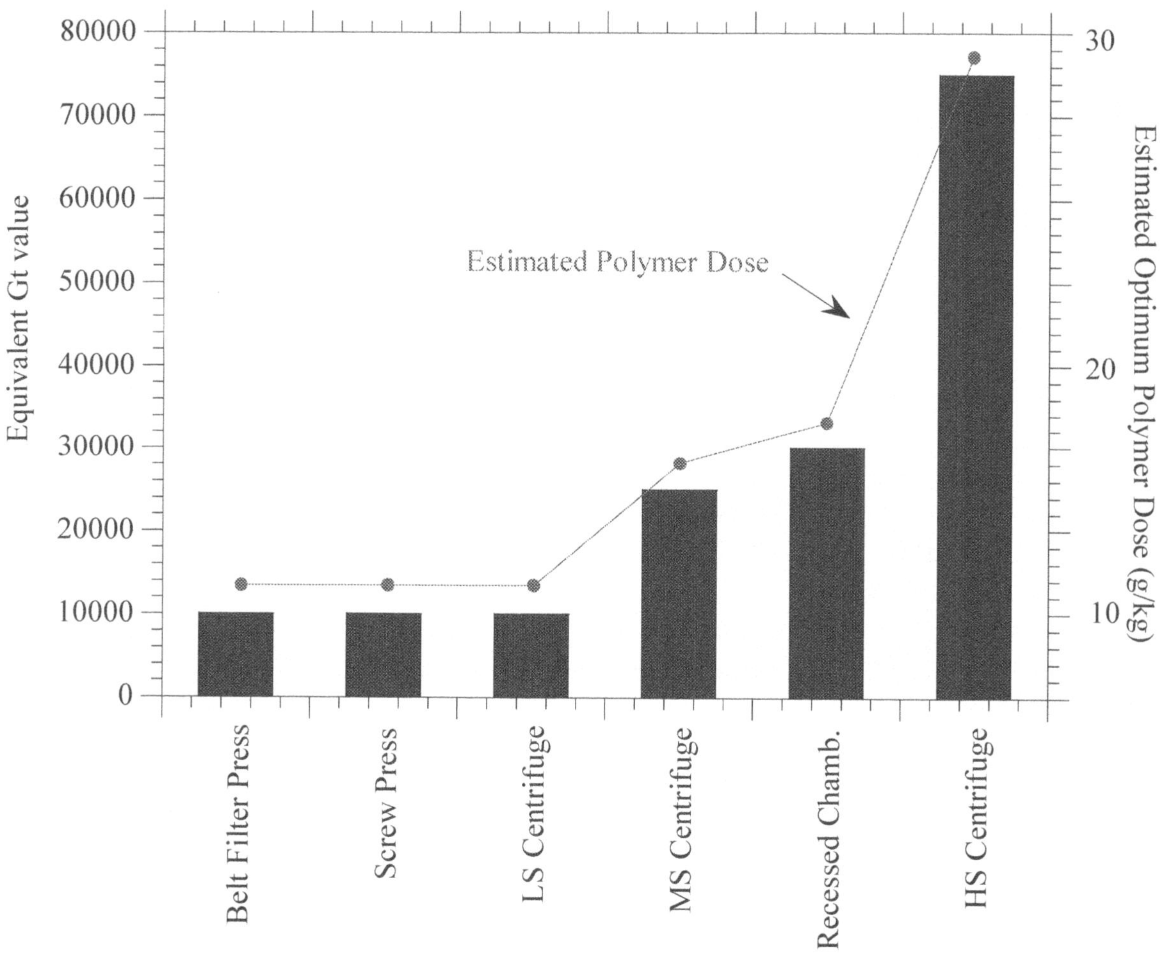

Figure 7-4. Estimated Polymer Demand for Anaerobically Digested Biosolids and Different Dewatering Equipment.

7.4 Conclusions and Implications

The development of an empirical model to predict polymer demand based on sludge and polymer characteristics and dewatering equipment produced reasonable results in terms of the correlation coefficient for the model. When the model was applied to a full-scale system to predict polymer demand, the model did a good job of predicting the OPD for a medium solids centrifuge, for two different sludge feeds, and two different polymer types. However, the model did not produce as good of a prediction for the high solids centrifuge at the plant, generally over-predicting the OPD by 33-85%. This may be related to the Gt or shear used for the high solids centrifuge overpredicting the actual shear during the trial. Based on the results, the model appears to hold some promise as a tool for predicting polymer demand under different scenarios. Additional research is needed to further refine the model, especially to develop a better understanding of shear during dewatering.

CHAPTER 8.0

RESEARCH APPLICATIONS: REDUCING POLYMER DEMAND

8.1 Overview

The results from this research and generation of a model that reasonably predicts polymer demand provides some insights into methods to decrease polymer demand. The three main parameters that have been shown to impact polymer demand are:

- Biocolloid concentration;
- Shear; and
- Polymer characteristics.

Using this knowledge, methods can potentially be developed to reduce polymer demand that focus on these factors. Although the concept of shear and its role in determining polymer demand has been around for several decades, the present research provides additional insights into the effects of different sludge types, different polymer types, and different dewatering technologies.

The following sections examine approaches that could be used to reduce polymer demand, and describe several case studies that demonstrate several of these approaches.

8.2 Reducing Biocolloid Concentration

Since the biocolloid concentration was directly linked to polymer demand, approaches that reduce the biocolloid concentration also would reduce the polymer demand. Biocolloids originate from microbial activity, and are primarily made up of protein and polysaccharides. Therefore, biocolloid concentration is affected by microbial activity, and these particles are also affected by different chemical constituents. The approaches for reducing biocolloid concentration fall into the following categories:

- Role of digestion
- Additives to reduce the biocolloids

8.2.1 Role of Digestion

Research from this study demonstrated a wide variation in biocolloid concentration for different digestion types. Thermophilic digestion processes resulted in much greater biocolloid concentrations than the mesophilic digestion processes, including both aerobic and anaerobic processes. Therefore, increased temperatures during digestion will lead to greater biocolloid concentration and subsequent polymer demand. In addition, this study found that the thermophilic sludges were more sensitive to shear, which can further increase polymer demand during conditioning and dewatering. Vogelaar et al. (2005) also reported that for laboratory scale anaerobic digestion of paper mill wastewater, thermophilic digestion had a greater content of colloidal sized (< 5 mm) particles compared to mesophilic digestion, and thermophilic digestion resulted in greater sensitivity to shear.

Houghton et al. (2000) found the concentration of thermal/solvent extractable exocellular biopolymer (ECP) was negatively correlated to the dewatering properties for mesophilic anaerobic digestion. However, the same study did not find any correlation between ECP and digester SRT. This study did not measure biocolloids in solution, but rather the concentration of biocolloid constituents (biopolymer) bound to the particles. Additional research is needed to better understand the role of digester SRT and polymer demand for conditioning.

A number of digestion enhancements have been suggested to improve digestion through increased volatile solids destruction. These enhancements generally rely on floc-disintegration techniques that will create a more degradable product. The release of more biocolloids, however, could increase polymer demand. For example, Kopp et al. (1997) reported that mechanical disintegration of sludge prior to digestion improved digestion but also increased the polymer demand for conditioning. They also found that for the control reactor with no pre-digestion enhancement, the polymer demand was relatively constant for SRTs of 3–15 days. Research into pre-digestion enhancements should consider the effects on dewatering and polymer demand since this will be an important component in fully understanding the impact of the process changes on solids treatment and disposal.

8.2.2 Chemicals and Amendments

Another possible method to reduce biocolloid concentration is through the addition of chemicals and amendments that enhance the degradation of biocolloids or bind/neutralize the biocolloids such that they do not consume polymer.

Addition of an enzyme mixture to sludges reduced the protein and polysaccharide content of two biosolids samples, and reduced the polymer demand and increased cake solids content (Ayol, 2005; Ayol and Dentel, 2005). This approach was cost effective (Ayol and Dentel, 2005), and appeared to be a viable tool for improving dewatering and reducing polymer demand. Full-scale testing is needed to further develop this concept. Neyens et al. (2004) reported that several advanced sludge treatment processes also were able to reduce the protein and polysaccharide concentration of sludges. Thermal hydrolysis, acid treatment and thermal hydrolysis, and peroxidation all reduced the protein and polysaccharide concentration of undigested thickened sludge. The study did not examine the effect on polymer demand, but research in the present study indicates that the reduction in protein and polysaccharide should reduce polymer demand. However, Neyens et al. (2004) examined "sludge" protein and polysaccharide and not just the biocolloid content. Additional research is needed to fully understand how these processes may impact biocolloids and polymer demand.

Biocolloid concentration also can be reduced through addition of chemicals, such as ferric chloride ($FeCl_3$) that sequester the biopolymer into the cake and reduce the demand associated with the biocolloids. For example, Murthy et al. (2000b) reported that addition of ferric chloride and alum was "extremely effective" at reducing polymer demand by removing biocolloidal material from solution. Similarly, Abu-Orf et al. (2001) reported relatively low dosages of ferric chloride (0.1 kg/kg DS) were able to reduce the polymer dosage by 50%. As part of this study, several experiments were performed to examine the effect of ferric chloride addition on polymer demand. The following sections exam two studies that investigate the impact of $FeCl_3$ addition both in the lab and in the field. The goals of these studies were, in part, to determine how polymer demand would be affected, and if potential cost savings would result.

The Philadelphia Water Department (PWD) has both high solids centrifuges (Humbolt) and medium solids centrifuges (Bird) that dewater two separate anaerobically digested sludges from their Northeast (NE) and Southwest (SW) plants. Laboratory experiments were performed to determine the effect of $FeCl_3$ addition on polymer demand, and this was followed up by a full-scale study on $FeCl_3$ addition. Another goal of the research undertaken by PWD was to reduce odor in the cake, although these results are not presented here. The results from both the laboratory and full-scale studies are presented below.

8.2.3 Laboratory Experiments with $FeCl_3$ Addition

A series of bench scale tests were used to determine the effect of iron and polymer addition on dewatering properties of the NE biosolids. The CST profile for $FeCl_3$ addition (Figure 8-1) is very similar to one for a typical polymer addition experiment; however, $FeCl_3$ alone does not produce superflocculated particles, and the particles are quite weak. As a result, polymer addition is also necessary for subsequent dewatering.

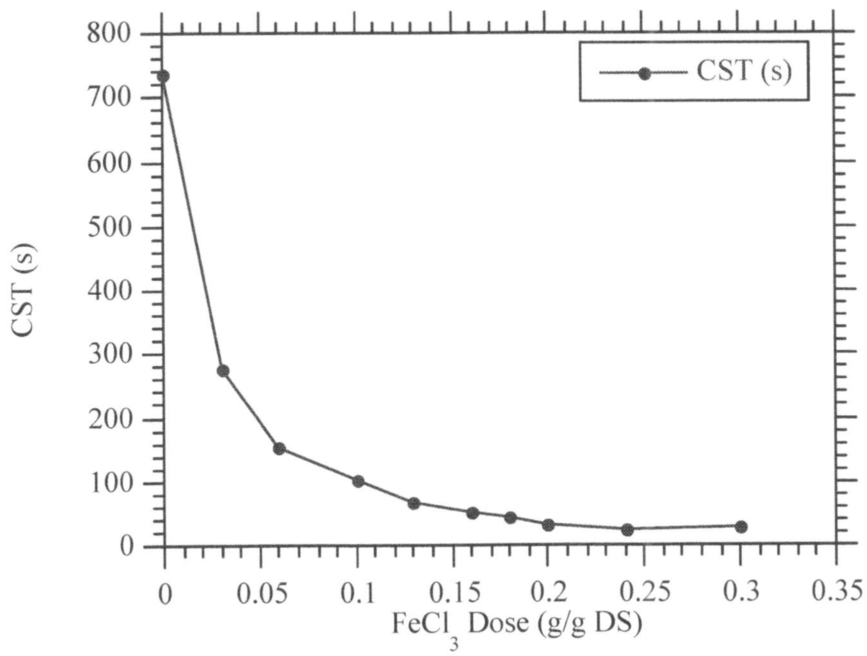

Figure 8-1. CST Profile for $FeCl_3$ Addition.

The effect of adding ferric chloride at three different doses followed by conditioning with cationic polymer is shown in Figure 8-2. Based on this analysis, the OPD as a function of the ferric chloride addition is shown in Figure 8-3. The results demonstrated that ferric chloride addition significantly reduced polymer demand. The OPD (measured with CST) was reduced by 46% when 0.1 kg of Fe was added per kg of biosolids on a dry basis. In addition, the concentration of soluble protein and SCOD decreased as the ferric chloride dose increased, and no significant change in the polysaccharide concentration was measured (see Figure 8-4). These results are consistent with those presented above and by Murthy et al. (2000b) in that the polymer demand is correlated strongly with the biocolloid concentration. Murthy et al. (2000b) examined the correlation between total biocolloid content after polymer addition but with no filtration. In the present study, the effect of polymer demand on the "soluble" biopolymer (that which passed the 4.2 μm filter) is similar, suggesting these fine particles can exert polymer demand. These preliminary results indicate that strategies could be developed to improve the dewaterability of the PWD biosolids and potentially reduce the polymer demand for conditioning and the associated operating costs for the biosolids processing facility. The addition of iron also might improve clarification, due to a sweep flocculation mechanism, and reduce polymer demand through charge neutralization.

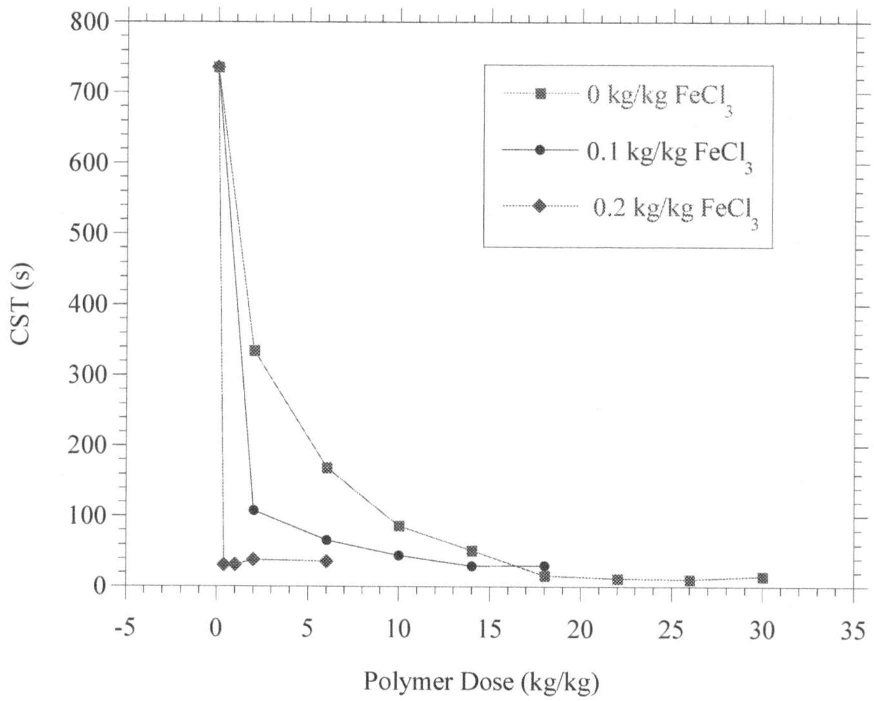

Figure 8-2. Effect of Ferric Chloride Dose on Polymer Dose for Conditioning.

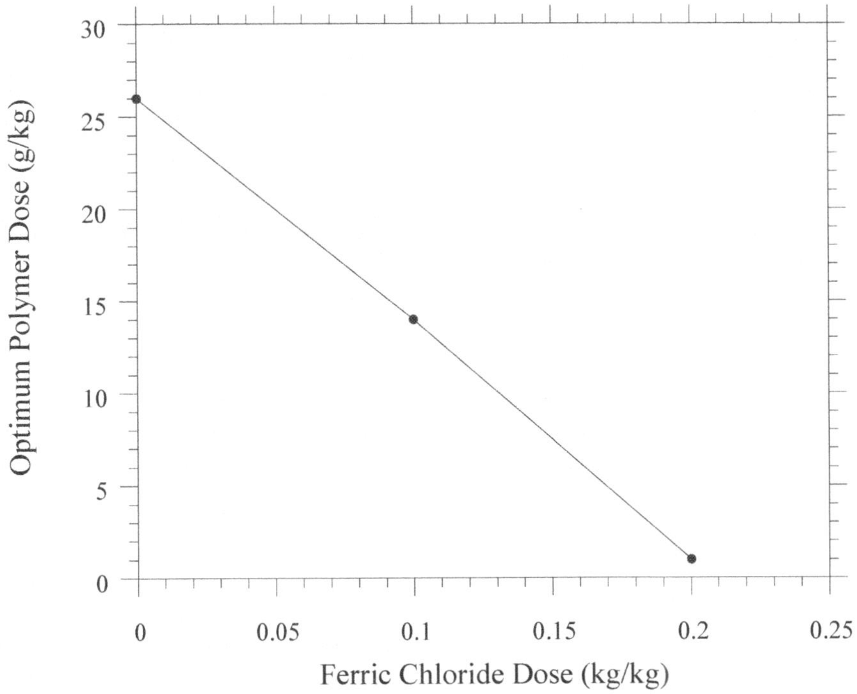

Figure 8-3. Effect of Ferric Chloride Addition on the Optimum Polymer Dose.

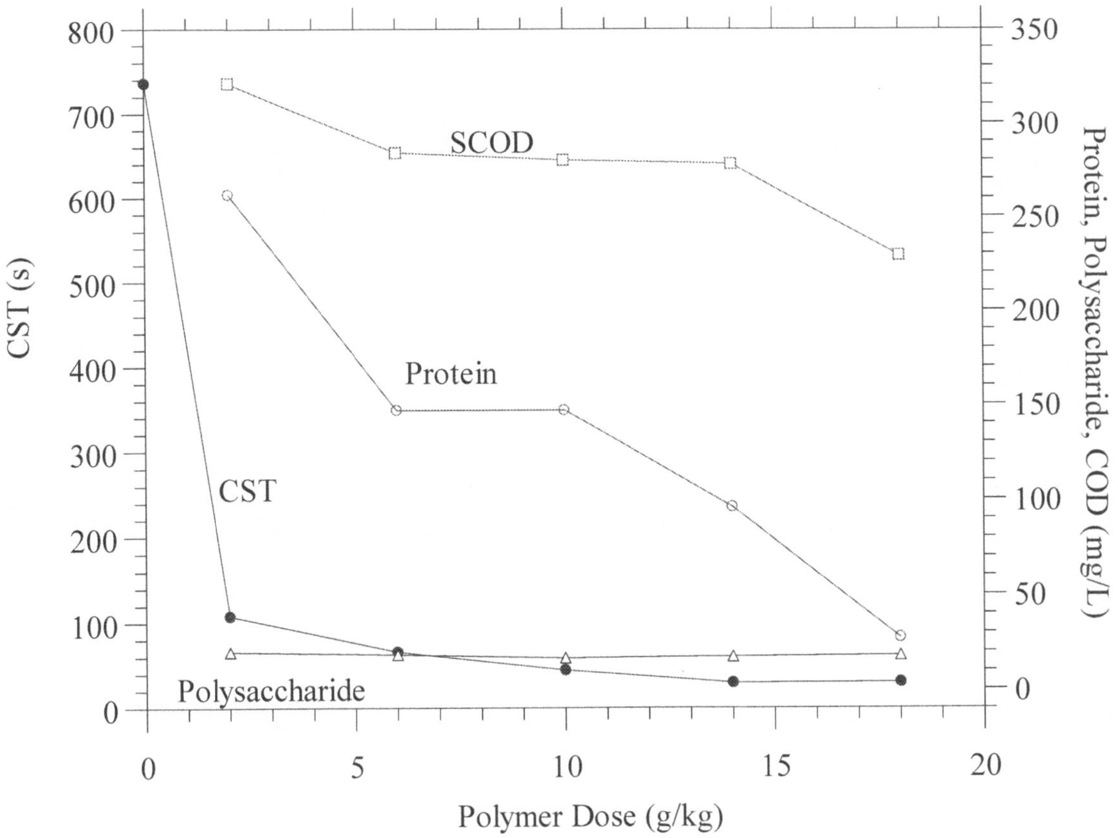

Figure 8-4. Effect of Ferric Chloride (0.1 kg/kg) and Polymer Dose on Soluble Biopolymer and SCOD.

8.2.4 Full-Scale Study with $FeCl_3$ Addition

Reductions in polymer demand measured in the laboratory experiments suggested that similar benefits might be seen in full-scale trials at the PWD Biosolids Processing Facility. The facility dewaters anaerobically digested biosolids using centrifuges. A trial was performed to examine the effect of ferric chloride addition on dewaterability measured by polymer demand, cake solids, and centrate quality. In this trial, two experiments were performed on the two different biosolids samples (NE and SW) using the same equipment operated under similar conditions.

The effect of ferric chloride addition on the OPD, pH, and the iron concentration measured in the cake is shown in Table 8-1. The addition of approximately 0.11 kg $FeCl_3$/kg DS resulted in a 32% reduction in polymer demand for both the NE and SW biosolids. The amount of iron measured in the cake solids after acid digestion and ICP analysis increased for both the NE and SW samples, although the SW sample had a greater increase in the iron content. The reasons for the better incorporation of the iron in the SW sample are unknown.

The effect of the ferric chloride addition on the cake total solids and the centrate quality is shown in Table 8-2. The cake total solids and cake volatile decreased in both the NE and SW samples with the iron addition. In addition, the percent volatile of the cake solids decreased with iron addition which would be expected with the greater iron content. The centrate TSS decreased by 57% and 30% in the NE and SW samples, respectively, while the volatile suspended solids (VSS) decreased by 68% and 16%, respectively. The centrate soluble protein decreased by 32% and 48% and the centrate total protein decreased by 51% and 26%, in the NE and SW samples, respectively.

Table 8-1. Effect of Ferric Chloride Addition on Polymer Demand.

Biosolids Sample	Iron Dose (kg/kg)	Fe in Biosolids (mg Fe/g DS)	pH after $FeCl_3$ addition	Optimum Polymer Dose (g/kg)
NE no iron	0	81.7	6.9	9.14
NE with iron	0.11	94.9	6.4	6.19
SW no iron	0	75.7	7.2	8.00
SW with iron	0.11	123.2	6.3	5.43

Table 8-2. Effect of Ferric Chloride Addition on Cake Solids and Centrate Quality.

Biosolids Sample	Iron Dose (kg/kg)	Total Cake Solids (% TS)	Centrate TSS (mg/L)	Centrate Soluble Protein (mg/L)	Centrate Total Protein (mg/L)
NE no iron	0	27.9	840	59.56	123.7
NE with iron	0.11	27.5	364	40.32	60.0
SW no iron	0	30.1	416	59.11	87.1
SW with iron	0.11	28.0	292	30.70	64.1

Based on an economic analysis which included cost of $FeCl_3$, polymer, cake hauling and disposal, the addition of $FeCl_3$ generally resulted in a slight increase to a slight savings which ranged from a savings of $3 per 1000 kg of DS to a cost of about $5.

The reduction in the centrate protein concentrations suggests that the cake protein content should be higher in samples with iron addition. However, the results showed that the cake protein in the iron-amended samples actually had less protein that could be extracted by distilled water extraction. This indicates that the iron plays a role in binding the biopolymer within the biofloc. This could have important implications for the production of odors during cake storage which are thought to result in part from degradation of proteins within the floc. In fact, additional testing showed iron addition reduced odors from the cake during storage. This could greatly impact costs if odors were limiting biosolids disposal options to more expensive alternatives such as land-filling.

8.2.5 Conclusions for Fe Addition

The results from this research demonstrated:

- Fe concentration in anaerobically digested biosolids was negatively correlated with optimum polymer dose (r^2=0.86).
- Addition of $FeCl_3$ prior to polymer addition reduced the polymer demand significantly in both laboratory and full-scale trials.
- The mechanism for reducing the polymer demand is through the sequestration of biocolloids into the solid phase.
- The results suggest iron addition could be an economical method for reducing polymer demand in difficult-to-dewater sludges.

8.3 Reducing Shear

Results from this research and others have shown that shear during conditioning and dewatering increases polymer demand. Approaches that reduce shear, therefore, will reduce the polymer demand. For example, reducing the shear associated with sludge transport through piping networks would help reduce polymer demand. It may also be possible to reduce shear within certain dewatering devices. For example, high solids centrifuges have been associated with high shear and increased polymer demand. Altering the location of polymer addition and the torque settings should reduce shear and lower polymer demand for these devices.

A trial was performed at the Western Lake Superior Sanitation District (WLSSD) to examine the role of shear on dewatering performance, including polymer demand. WLSSD uses a TPAD process for digestion followed by high solids centrifugation for dewatering. The trials for this study examined the impact of adjusting the centrifuge torque, as well as examining different polymer addition points, on dewatering performance. In each case, the purpose was to examine the effects of shear on polymer demand, and ultimately develop a dewatering strategy to reduce shear and thereby decrease costs. The results from the two studies are provided in the following two sections.

8.3.1 Effect of Polymer Addition Point

In this study, WLSSD and the research team examined the effect of polymer addition location on subsequent polymer demand. High solids centrifuges generally have high polymer demands as a result of high shear, as shown in the previous chapter. Much of the shear is thought to occur in the entrance zone of the centrifuge and possibly near the exit zone of the centrifuge. WLSSD has the ability to add polymer at three different locations. One location is directly into the centrifuge, the second is in the feed pipe approximately 0.3 m upstream of the centrifuge, and the third location is in the feed pipe approximately 20 m upstream of the centrifuge. These three locations should result in different amounts of shear and therefore different polymer requirements.

A study was performed in which the three polymer addition points were used, and for each location, the polymer dose was varied to obtain the optimum polymer dose based on the cake and centrate quality. During the study, all other parameters (such as torque and throughput) were maintained constant.

The effect of polymer addition location on the subsequent polymer demand and dewatering performance is summarized in Table 8-3. The impact of polymer addition location on the OPD showed the further upstream, the greater the polymer demand. Other researchers have shown that shear occurs in the piping network entering the dewatering device, and the greater amount of piping (or headloss due to piping), the greater the shear and subsequent polymer dose (Murthy et al., 1997). This study showed similar results. The Gt for each location was determined using the previously described technique. Adding the polymer 20 m upstream resulted in a greater than 50% increase in the Gt. Most interesting is that adding the polymer directly into the centrifuge resulted in a relatively low shear or Gt, about 11,250 (see Table 8-3). Adding the polymer in this location also provided very similar cake solids content and centrate quality, so dewatering performance was very similar overall. This is a very low Gt value for a high solids centrifuge compared to other systems tested, for which results are provided in Chapter 6.0. Adding the polymer immediately upstream of the centrifuge doubled the shear or associated Gt, suggesting that the entrance zone is a relatively high shear zone in the centrifuge.

Table 8-3. Summary of Results from Field Trial with Different Polymer Addition Locations.

Polymer Addition Location	Polymer Dose (g/kg)	Centrate Soluble Protein (mg/L)	Centrate Soluble Polysaccharides (mg/L)	Cake Solids (%)	Gt Value from Shear Tests
In the centrifuge	44.0	307	86.6	29.72	11250
0.3 m upstream of centrifuge	49.8	320	93.1	30.24	21000
20 m upstream of centrifuge	53.4	293	86.2	30.14	34500

These results demonstrate that addition of the polymer directly into the centrifuge can significantly reduce shear and polymer demand, while not significantly sacrificing dewatering performance, as measured by cake solids and centrate quality. WLSSD uses polymer addition directly into the centrifuge in order to achieve these benefits. The odor measured as dilutions-to-threshold also doubled in cake samples when comparing the addition point within the centrifuge and just upstream of the centrifuge. These results demonstrate the role of shear in odor as well as polymer demand.

8.3.2 Effect of Centrifuge Torque

In this study, four different centrifuge torques were used (holding all other parameters constant), and the OPD was determined for each torque setting. The torque settings were 40%, 50%, 65%, and 80%. A summary of the results from this study are provided in Table 8-4. The OPD increased proportionally with the torque setting, and the cake solids also increased with the greater torque setting. The Gt measured for each torque setting also increased linearly with the OPD. The greater torque and OPD resulted in improved centrate quality, although all the parameters were within acceptable ranges for normal operations. The torque settings showed shear could be increased (or decreased) by a factor of about three by changing the centrifuge parameters. Based on these results, WLSSD now operates the centrifuge at a 40% torque (previously had used 50%), to reduce polymer costs. The plant also reduced throughput to achieve the desired cake solids. Another benefit is a reduction in odor at the lower torque. Interestingly, odor was also an important factor in the decision making process for determining operating procedures.

Table 8-4. Summary of Data for Torque Setting Study.

Torque Setting	Polymer Dose (g/kg)	Centrate Soluble Protein (mg/L)	Centrate Soluble Polysaccharides (mg/L)	Cake Solids (%)	Gt Value from Shear Tests
40% load	40.9	365	90.1	28.0	7500
50% load	44.0	307	86.6	29.7	11250
65% load	46.8	293	83.4	31.4	18750
80% load	49.3	274	81.8	31.7	21000

8.3.3 Conclusions for Polymer Addition Location and Torque Studies

These full-scale studies support the laboratory findings that shear is an important parameter for determining polymer dose, and confirm that operations can be adjusted to minimize shear. For example, the addition of polymer should occur at locations that minimize the shear through piping systems as well as the entrance zone of a centrifuge. Similarly, reducing torque can reduce polymer demand, although there may be a tradeoff with cake solids content. Plants will have to examine their particular situation and balance polymer costs with disposal costs to determine the most cost effective operational parameters.

8.4 Polymer Characteristics and Dosing

Another avenue for reducing polymer demand is through selection of the polymer and control of polymer dosing. Research on polymer dosing control has focused on maintaining near-OPD through feedback control in order to maintain a consistent product and also minimize polymer dosage. Advances in sensor technology and algorithms have made automation a practical and useful tool for utilities (e.g., see Dentel and Abu-Orf, 1995; Dentel et al., 2000; Abu-Orf and Dentel, 1998; Brady and Montoya, 2005). Ormeci and Abu-Orf (2005) reported the use of network strength measurements as a method to choose polymer types and control polymer dosages in order to match the shear that occurs during dewatering. This would create flocs with network strength such that water could be released and better dewatering could be achieved.

When selecting polymers, the least expensive polymer should be chosen to achieve the desired product quality in terms of cake solids and filtrate/centrate solids. Research in this study and others indicates that high MW polymers are needed to adequately condition sludges. In addition, structured polymers may perform better under high shear situations (Weir and Moody, 2003; Pearse et al., 2001). Therefore, in a treatment scenario with high shear dewatering devices (such as a high solids centrifuge) and shear-sensitive sludges, structured polymers may be a good choice as a conditioning chemical.

Several reports examined the use of dual polymer conditioning, where one polymer type (such as anionic, non-ionic, or cationic polymer) is added, followed by a second polymer with different properties. Lee and Liu (2000) found that dual polymer conditioning with a non-ionic and cationic polymer improved the dewatering properties of a waste activated sludge, but did not reduce the total polymer demand. Senthilnathan and Sigler (1993) also found that dewatering properties improved, including cake solids and filtrate quality on both laboratory filtration tests and a full-scale belt filter press, however, the polymer dose was not reduced. Fan et al. (2000) found that flocculation of alumina was enhanced using a low MW anionic polymer followed by a high MW cationic polymer. The use of anionic polymer along with cationic polymer to condition an ATAD sludge improved dewatering properties and decreased the cationic polymer demand (Agarwal et al., 2005). These results show that dual polymer conditioning may be an important method for improving dewatering properties, although it may not reduce the overall polymer demand.

CHAPTER 9.0

SUMMARY AND ADDITIONAL RESEARCH NEEDS

9.1 Overview

The research undertaken in this project sought to better understand the factors which impact polymer demand for conditioning. The research focused on sludge and polymer properties, as well as shear, and attempted to understand the interactions of these different parameters. Over the last few decades, research has been performed to understand the role of shear and the impact of polymer characteristics; however, none sought to combine all these factors into a single model to estimate polymer demand. This study provided a relatively large data set using consistent testing protocols to examine the different factors independently. The empirical equation that was developed is a starting point for attaining the goal of an accurate, universal model to predict polymer demand. It also provides an analysis tool to develop methods to reduce polymer demand.

9.2 Future Research Needs

Difficulties that were encountered in this research included quantifying the effect of shear. The research showed that shear during conditioning (liquid shear) was different than shear during dewatering (cake shear). Most researchers have focused on liquid shear. However, the results of this research indicate that shear on the cake may have a bigger impact on cake quality in terms of odor production compared to shear of the liquid during conditioning. If the shear experienced by the cake can be replicated in the laboratory, the relationship between cake shear and polymer demand could be developed.

Estimating or replicating cake shear would provide a more realistic approach to understanding variables that impact floc properties, polymer demand and product quality. Methods to quantify the shear on cake during dewatering would allow plant operators to predict and minimize polymer demand during conditioning and dewatering, while maintaining product quality (cake solids and cake odor characteristics).

REFERENCES

Abu–Orf, M.M. and S.K. Dentel. 1997. Polymer dose assessment using the streaming current detector. *Water Environ. Res.*, **69**, 1075-1085.

Abu–Orf, M.M. and S.K. Dentel. 1998. Automatic control of polymer dose using the streaming current detector. *Water Environ. Res.*, **70**, 1005-1018.

Abu-Orf, M.M., P. Griffin, and S.K. Dentel. 2001. Chemical and Physical Pretreatment of Autothermal Thermophylic Aerobically Digested (ATAD) Biosolids for Dewatering. *Proc. 74th Annu. Conf. Water Environ. Fed.*, Atlanta, GA.

Agarwal, S., M. Abu-Orf, and J.T. Novak. 2005. Sequential polymer dosing for effective dewatering of ATAD sludges. *Wat. Res.*, **39**, 1301-1310.

American Public Health Association (APHA), American Water Works Association (AWWA), and Water Environment Federation (WEF). 1995. Standard Methods for the Examination of Water and Wastewater.

Ayol, A. 2005. Enzymatic treatment effects on dewaterability of anaerobically digested biosolids-I: performance evaluations. *Process Biochem.*, **40**, 2427-2434.

Ayol, A. and S.K. Dentel. 2005. Enzymatic treatment effects on dewaterability of anaerobically digested biosolids-II: laboratory characterization of drainability and filterability. *Process Biochem.*, **40**, 2435-2442.

Brady, P. and M. Montoya. 2005. Control Technologies. *Proceedings of WEF/AWWA Joint Residuals and Biosolids Management Conference*. Nashville, TN.

Brown, M.J. and J.N. Lester. 1980. Comparison of Bacterial Extracellular Polymer Extraction Methods. *App. Environ. Microbiol.*, **40**, 179-185.

Dentel, S.K. and M.M. Abu-Orf. 1995. Laboratory and Full-Scale Studies of Liquid Stream Viscosity and Streaming Current for Characterization and Monitoring of Dewaterability. *Wat. Res.*, **29**, 2663-2672.

Dentel, S.K., M. Abu-Orf, and C.A. Walker. 2000. Optimization of slurry flocculation and dewatering based on electrokinetic and rheological phenomena. *Chemical Engineering Journal*, **80**, 65-72.

Dentel, S.K. 2001. Conditioning. In *Sludge Into Biosolids, Processing, Disposal, and Utilization.* Eds. L. Spinosa and P.A. Vesilind. IWA Publishing, London.

Dentel S.K., M.M. Abu-Orf, and N.J. Griskowitz. 1993. *Guidance Manual for Polymer Selection in Wastewater Treatment Plants*. Alexandria, VA: Water Environment Research Foundation (Publication D0013).

Dignac, M.F., V. Urbain, D. Rybacki, A. Bruchet, D. Snidaro, and P. Scribe. 1998. Chemical Description of Extracellular Polymers: Implication on Activated Sludge Floc Structure. *Wat. Sci. Technol.*, **38**, 39-46.

Dubois, M., K.A. Gilles, J.K. Hamilton, P.A. Rebers, and F. Smith. 1956. Colorimetric Method for Determination of Sugars and Related Substances. *Anal. Chem.*, **28**, 350-357.

Fan, A., N.J. Turro, and P. Somasundaran. 2000. A study of dual polymer flocculation. *Coll. Surfaces*, **162**, 141-148.

Feitz, A.J., J. Guan, and T.D. Waite. 2002. Effect of iron on anaerobic sludge and its dewaterability. *Third International Water Association Congress*, Melbourne, April 2002.

Forster, C.F. and D.C. Lewin. 1972. Polymer Interactions at Activated Sludge Surfaces. *J. Effl. Wat. Trt.* **12**, 520-525.

Frolund, B., R. Palmgren, K. Keiding, and P.H. Nielsen. 1996. Extraction of extracellular polymers from activated sludge using a cation exchange resin. *Wat. Res.*, **30**, 1749-1758.

Galla, C.A., D.L. Freedman, B.F. Severin, and B.J. Kim. 1999. Laboratory Prediction of Belt Filter Press Dewatering Dynamics. *Proc. 72nd Ann. Conf. Water Environ. Fed..*, Orlando, FL.

Gregory, J. 1993. The Role of Colloid Interactions in Solid-Liquid Separation. *Wat. Sci. Tech.*, **27**, 1-17.

Hartree, E.F. 1972. Determination of Protein: A Modification of the Lowry Method That Gives a Linear Photometric Response. *Anal. Biochem.*, **48**, 422-427.

Higgins, M.J. and J.T. Novak. 1997a. Characterization of Exocellular Protein and Its Role in Bioflocculation. *J. of Env. Eng.*, **123**, 479-485.

Higgins, M.J. and J.T. Novak. 1997b. The Effect of Cations on the Settling and Dewatering of Activated Sludges: Laboratory Results. *Water Environ. Res.*, **69**, 215-224.

Higgins, M.J., K. Hamel, Y.C. Chen, S.N. Murthy, E.J. Barben, A. Livadaros, M. Travis, and N.A. Maas. 2005a. Part II of Field Research: Impact of Centrifuge Torque and Polymer Dose on Odor Production from Anaerobically Digested Biosolids. *Proceedings of WEF/AWWA Joint Residuals and Biosolids Management Conference*. Nashville, TN.

Higgins, M.J., D.P. Yarosz, Y.C. Chen, S.N. Murthy, N. Maas, J. Cooney, D. Glindemann, and J.T. Novak. 2005b. Cycling of volatile organic sulfur compounds in anaerobically digested biosolids and its implications for odors. In Press, *Water Env. Research.*

Horan, N.J. and C.R. Eccles. 1986. Purification and Characterization of Extracellular Polysaccharide from Activated Sludges. *Wat. Res.*, **20**, 1427-1432.

Houghton, J.I., T. Stephenson, and J. Quarmby. 2000. The impact of digester retention time on microbial extracellular polymer production and sludge dewaterability. *Env. Technol.,* **22**, 1157-1165.

Kam, S.K., and J. Gregory. 1999. Charge determination of synthetic cationic polyelectrolytes by colloid titration. *Colloids and Surfaces,* **159**, 165-179.

Karr, P.R. and T.M. Keinath. 1978. Influence of particle size on sludge dewaterability. *Jour. Wat. Pollut. Control Fed.,* **50**, 1911-1929.

Kopp, J., J. Muller, N. Dichtl, and J. Schwedes. 1997. Anaerobic Digestion and Dewatering Characteristics of Mechanically Disintegrated Excess Sludge. *Wat. Sci. Technol.*, **36**, 129-136.

Lee, C.H. and J.C. Liu. 2000. Enhanced Sludge Dewatering by Dual Polyelectrolytes Conditioning. *Wat. Res.*, **34**, 4430-4436.

Lowry, O.H., N.J. Rosebrough, A.L. Farr, and R.J. Randall. 1951. Protein Measurement with the Folin Phenol Reagent. *J. Biol. Chem.*, **193**, 265-275.

Murthy, S.N., J.T. Novak, and M. Buckley. 1997. Predicting polymer conditioning requirements in high pressure sludge dewatering devices. *Mid-Atlantic Industrial and Hazardous Waste,* **29**, 293-302.

Murthy, S.N. and J.T. Novak. 1999. Factors affecting floc properties during aerobic digestion: Implications for dewatering. *Water Envir. Res.,* **71**, 197-202.

Murthy, S.N., R.D. Holbrook, J.T. Novak, and F. Surovik. 2000a. Mesophilic Aeration of ATAD sludge to improve plant operations. *Water Envir. Res.,* **72**, 476-483.

Murthy, S.N., J.T. Novak, and R.D. Holbrook. 2000b. Optimizing Dewatering of Biosolids from Autothermal Thermophilic Aerobic Digesters (ATAD) Using Inorganic Conditioners. *Water Envir. Res.,* **72**, 714-721.

Neyens, E., J. Baeyens, R. Dewil, and B. De Heyder. 2004. Advanced sludge treatment affects extracellular polymeric substances to improve activated sludge dewatering. *J. Haz. Mater.*, **106B**, 83-92.

Nielsen, P.H. and K. Keiding. 1998. Disintegration of activated sludge flocs in the presence of sulfide. *Wat. Res.*, **32**, 313-320.

Novak, J.T., J.F. Prendeville, and J.H. Sherrard. 1988. Mixing Intensity and Polymer Performance in Sludge Dewatering. *J. Env. Eng.*, **114**, 190-198.

Novak, J.T. and N. Bandak. 1989. Chemical conditioning and the resistance of sludges to shear. *Water Envir. Res.,* **61**, 327-332.

Novak, J.T. and D.P. Lynch. 1990. The Effect of Shear on Conditioning: Chemical Requirements During Mechanical Sludge Dewatering. *Wat. Sci. Tech.*, **22**, 117-124.

Novak, J., W. Knocke, W. Burgos, and P. Schuler. 1993. Predicting the Dewatering Performance of Belt Filter Presses. *Wat. Sci. Tech.*, **28**, 11-19.

Novak, J.T., M. Sadler, and S.N. Murthy. 1999. Mechanisms influencing conditioning and dewatering of aerobically and anaerobically digested sludge. *Proc. 72nd Ann. Conf. Water Envir. Fed..*, Orlando, FL.

Ormeci, B. and M.M. Abu-Orf. 2005. Protocol to Measure Network Strength of Sludges and Its Implications to Dewatering. *J. Env. Eng.*, **131**, 80-85.

Pearse, M.J., S. Weir, S.J. Adkins, and G.M. Moody. 2001. Advances in Mineral Flocculation. *Min. Eng.*, **14**, 1505-1511.

Roberts, K. and O. Olsson. 1975. Influence of Colloidal Particles on Dewatering of Activated Sludge with Polyelectrolyte. *Env. Sci. Technol.*, **9**, 945-948.

Senthilnathan, P.R. and R.G. Sigler. 1993. Improved Sludge Dewatering by Dual Polymer Conditioning. *Wat. Sci. Tech.*, **28**, 53-57.

Standard Methods for the Examination of Water and Wastewater. 1999. 18th ed., Am. Public Health Assoc., Washington, D.C.

Vogelaar, J.C.T., A. De Keizer, S. Spijker and G. Lettinga. 2005. Bioflocculation of mesophilic and thermophilic activated sludge, *Water Research*, **39**, 37-46.

Weir, S. and G.M. Moody. 2003. The importance of flocculant choice with consideration to mixing energy to achieve efficient solid/liquid separation. *Min. Eng.*, **16**, 109-113.

Werle, C.P., J.T. Novak, W.R. Knocke, and J.H. Sherrard. 1984. Mixing Intensity and Polymer Sludge Conditioning. *J. Env. Eng.*, **110**, 919-934.

WASTEWATER UTILITY

Alabama
Montgomery Water Works & Sanitary Sewer Board

Alaska
Anchorage Water & Wastewater Utility

Arizona
Gila Resources
Glendale, City of, Utilities Department
Mesa, City of
Peoria, City of
Phoenix, City of
Pima County Wastewater Management

Arkansas
Little Rock Wastewater Utility

California
Calaveras County Water District
Central Contra Costa Sanitary District
Corona, City of
Crestline Sanitation District
Delta Diablo Sanitation District
Dublin San Ramon Services District
East Bay Dischargers Authority
East Bay Municipal Utility District
Eastern Municipal Water District
El Dorado Irrigation District
Fairfield-Suisun Sewer District
Fresno Department of Public Utilities
Inland Empire Utilities Agency
Irvine Ranch Water District
Las Virgenes Municipal Water District
Livermore, City of
Lodi, City of
Los Angeles, City of
Los Angeles County, Sanitation Districts of
Napa Sanitation District
Orange County Sanitation District
Palo Alto, City of
Riverside, City of
Sacramento Regional County Sanitation District
San Diego Metropolitan Wastewater Department, City of
San Francisco, City & County of
San Jose, City of
Santa Barbara, City of
Santa Cruz, City of
Santa Rosa, City of
South Bayside System Authority
South Coast Water District
South Orange County Wastewater Authority
Stege Sanitary District
Sunnyvale, City of
Union Sanitary District
West Valley Sanitation District

Colorado
Aurora, City of
Boulder, City of
Colorado Springs Utilities
Greeley, City of
Littleton/Englewood Water Pollution Control Plant
Metro Wastewater Reclamation District, Denver

Connecticut
The Mattabassett District
New Haven, City of, WPCA

District of Columbia
District of Columbia Water & Sewer Authority

Florida
Broward, County of
Fort Lauderdale, City of
JEA
Miami-Dade Water & Sewer Authority
Orange County Utilities Department
Reedy Creek Improvement District
Seminole County Environmental Services
St. Petersburg, City of
Stuart Public Utilities
Tallahassee, City of
Tampa, City of
Toho Water Authority
West Palm Beach, City of

Georgia
Atlanta Department of Watershed Management
Augusta, City of
Clayton County Water Authority
Cobb County Water System
Columbus Water Works
Fulton County
Gwinnett County Department of Public Utilities
Savannah, City of

Hawaii
Honolulu, City & County of

Idaho
Boise, City of

Illinois
American Bottoms Wastewater Treatment Plant
Greater Peoria Sanitary District
Kankakee River Metropolitan Agency
Metropolitan Water Reclamation District of Greater Chicago
Wheaton Sanitary District

Iowa
Ames, City of
Cedar Rapids Wastewater Facility
Des Moines Metro Wastewater Reclamation Authority
Iowa City

Kansas
Johnson County Unified Wastewater Districts
Unified Government of Wyandotte County/ Kansas City, City of

Kentucky
Louisville & Jefferson County Metropolitan Sewer District

Louisiana
Sewerage & Water Board of New Orleans

Maine
Bangor, City of
Portland Water District

Maryland
Anne Arundel County Bureau of Utility Operations
Hagerstown, City of
Howard County Department of Public Works
Washington Suburban Sanitary Commission

Massachusetts
Boston Water & Sewer Commission
Upper Blackstone Water Pollution Abatement District

Michigan
Ann Arbor, City of
Detroit, City of
Holland Board of Public Works
Lansing, City of
Saginaw, City of
Wayne County Department of Environment
Wyoming, City of

Minnesota
Rochester, City of
Western Lake Superior Sanitary District

Missouri
Independence, City of
Kansas City Missouri Water Services Department
Little Blue Valley Sewer District
Metropolitan St. Louis Sewer District

Nebraska
Lincoln Wastewater System

Nevada
Henderson, City of
Reno, City of

New Jersey
Bergen County Utilities Authority
Ocean, County of
Passaic Valley Sewerage Commissioners

New York
New York City Department of Environmental Protection

North Carolina
Charlotte/Mecklenburg Utilities
Durham, City of
Metropolitan Sewerage District of Buncombe County
Orange Water & Sewer Authority

Ohio
Akron, City of
Butler County Department of Environmental Services
Columbus, City of
Metropolitan Sewer District of Greater Cincinnati
Northeast Ohio Regional Sewer District
Summit, County of

Oklahoma
Tulsa, City of

Oregon
Clean Water Services
Eugene/Springfield Water Pollution Control
Water Environment Services

Pennsylvania
Philadelphia, City of
University Area Joint Authority

South Carolina
Charleston Commissioners of Public Works
Mount Pleasant Waterworks & Sewer Commission
Spartanburg Sanitary Sewer District

Tennessee
Cleveland, City of
Knoxville Utilities Board
Murfreesboro Water & Sewer Department
Nashville Metro Water Services

Texas
Austin, City of
Dallas Water Utilities
Denton, City of
El Paso Water Utilities
Fort Worth, City of
Gulf Coast Waste Disposal Authority
Houston, City of
San Antonio Water System
Trinity River Authority

Utah
Salt Lake City Corporation

WERF SUBSCRIBERS

Virginia
Alexandria Sanitation Authority
Arlington, County of
Fairfax County
Hampton Roads Sanitation District
Henrico, County of
Hopewell Regional Wastewater Treatment Facility
Loudoun County Sanitation Authority
Lynchburg Regional WWTP
Prince William County Service Authority
Richmond, City of
Rivanna Water & Sewer Authority

Washington
Edmonds, City of
Everett, City of
King County Department of Natural Resources
Seattle Public Utilities
Sunnyside, Port of
Yakima, City of

Wisconsin
Green Bay Metro Sewerage District
Kenosha Water Utility
Madison Metropolitan Sewerage District
Milwaukee Metropolitan Sewerage District
Racine, City of
Sheboygan Regional Wastewater Treatment
Wausau Water Works

Australia
South Australian Water Corporation
Sydney Water Corporation
Water Corporation of Western Australia

Canada
Greater Vancouver Regional District
Regina, City of, Saskatchewan
Toronto, City of, Ontario
Winnipeg, City of, Manitoba

Mexico
Servicios de Agua y Drenaje de Monterrey, I.P.D.

New Zealand
Watercare Services Limited

United Kingdom
Yorkshire Water Services Limited

STORMWATER UTILITY

California
Los Angeles, City of, Department of Public Works
Monterey, City of
San Francisco, City & County of
Santa Rosa, City of

Colorado
Aurora, City of
Boulder, City of

Georgia
Griffin, City of

Iowa
Cedar Rapids Wastewater Facility
Des Moines Metro Wastewater Reclamation Authority

Kansas
Overland Park, City of

Maine
Portland Water District

Minnesota
Western Lake Superior Sanitary District

North Carolina
Charlotte, City of, Stormwater Services

Pennsylvania
Philadelphia, City of

Tennessee
Chattanooga Stormwater Management

Washington
Bellevue Utilities Department
Seattle Public Utilities

STATE

Arkansas Department of Environmental Quality
Fresno Metropolitan Flood Control District
Kansas Department of Health & Environment
Kentucky Department of Environmental Protection
Ohio River Valley Sanitation Commission
Urban Drainage & Flood Control District, CO

CORPORATE

Abt. Associates
ADS Environmental Services
The ADVENT Group Inc.
Alan Plummer & Associates
Alden Research Laboratory Inc.
American Water
Aqua-Aerobic Systems Inc.
Aquateam–Norwegian Water Technology Centre A/S
ARCADIS
Associated Engineering
Black & Veatch
Blasland, Bouck & Lee Inc.
Boyle Engineering Corporation
Brown & Caldwell
Burns & McDonnell
CABE Associates Inc.
The Cadmus Group
Camp Dresser & McKee Inc.
Carollo Engineers Inc.
Carpenter Environmental Associates Inc.
CDS Technologies Inc.
Chemtrac Systems Inc.
CH2M HILL
Construction & Environmental Consulting, Inc.
Damon S. Williams Associates, LLC
David L. Sheridan, P.C.
Dewling Associates Inc.
Earth Tech Inc.
Ecovation
EMA Inc.
The Eshelman Company Inc.
EWT Holdings Corporation
Fay, Spofford, & Thorndike Inc.
Freese & Nichols Inc.
ftn Associates Inc.
Fuss & O'Neill Inc.
Gannett Fleming Inc.
Geosyntec Consultants
GHD
Greeley and Hansen LLC
Hazen & Sawyer, P.C.
HDR Engineering Inc.
HNTB Corporation
HydroQual Inc.
Infilco Degremont Inc.
Institute for Environmental Technology & Industry, Korea
Jacobson Helgoth Consultants Inc.
Jason Consultants LLC Inc.
Jordan, Jones, & Goulding Inc.
KCI Technologies Inc.
Kelly & Weaver, P.C.
Kennedy/Jenks Consultants
KMK Consultants
Komline Sanderson Engineering Corporation
Lawler, Matusky & Skelly Engineers, LLP
Limno-Tech Inc.
LTI Inc.
Lombardo Associates Inc.
Malcolm Pirnie Inc.
Material Matters
McKim & Creed
MEC Analytical Systems Inc.
Metcalf & Eddy Inc.
MWH
New England Organics
O'Brien & Gere Engineers Inc.
Odor & Corrosion Technology Consultants Inc.
Oswald Green, LLC
PA Government Services Inc.
Parametrix Inc.
Parsons
Post, Buckley, Schuh & Jernigan
R&D Engineering/Conestoga Rover & Associates
RMC Inc.
R.M. Towill Corporation
Ross & Associates Ltd.
Rothberg, Tamburini & Windsor, Inc.
Royce Technologies
SAIC Maritime Technical Group
Stantec Consulting Inc.
Stearns & Wheler, LLC
Stone Environmental Inc.
Stormwater360
Stratus Consulting Inc.
Synagro Technologies Inc.
Tetra Tech Inc.
Trojan Technologies Inc.
URS Corporation
USfilter
Wade-Trim Inc.
Westin Engineering Inc.
Weston Solutions Inc.
Woodard & Curran
WRc/D&B, LLC
WWETCO, LLC
Zenon Environmental Inc.
Zoeller Pump Company

INDUSTRY

American Electric Power
ChevronTexaco Energy Research & Technology Company
The Coca-Cola Company
Dow Chemical Company
DuPont Company
Eastman Chemical Company
Eastman Kodak Company
Eli Lilly & Company
Merck & Company Inc.
Procter & Gamble Company
RWE Thames Water Plc
Severn Trent Services Inc.
Suez Environnment
United Water Services LLC

Note: List as of 10/1/05

WERF Product Order Form

As a benefit of joining the Water Environment Research Foundation, subscribers are entitled to receive one complimentary copy of all final reports and other products. Additional copies are available at cost (usually $10). To order your complimentary copy of a report, please write "free" in the unit price column. WERF keeps track of all orders. If the charge differs from what is shown here, we will call to confirm the total before processing.

Name ________________ Title ________________

Organization ________________

Address ________________

City ________ State ________ Zip Code ________ Country ________

Phone ________ Fax ________ Email ________

Stock #	Product	Quantity	Unit Price	Total
			Postage & Handling	
			VA Residents Add 4.5% Sales Tax	
			Canadian Residents Add 7% GST	
			TOTAL	

Method of Payment:

(All orders must be prepaid.)

- ❑ Check or Money Order Enclosed
- ❑ Visa ❑ Mastercard ❑ American Express

Account No. ________________ Exp. Date ________

Signature ________________

Shipping & Handling:

Amount of Order	United States	Canada & Mexico	All Others
Up to but not more than:	Add:	Add:	Add:
$20.00	$5.00*	$8.00	50% of amount
30.00	5.50	8.00	40% of amount
40.00	6.00	8.00	
50.00	6.50	14.00	
60.00	7.00	14.00	
80.00	8.00	14.00	
100.00	10.00	21.00	
150.00	12.50	28.00	
200.00	15.00	35.00	
More than $200.00	Add 20% of order	Add 20% of order	

*minimum amount for all orders

Note: Please make checks payable to the Water Environment Research Foundation.

To Order (Subscribers Only):

Log on to www.werf.org and click on the "Product Catalog."

Phone: (703) 684-2470
Fax: (703) 299-0742.

WERF
Attn: Subscriber Services
635 Slaters Lane
Alexandria, VA 22314-1177

To Order (Non-Subscribers):

Non-subscribers may be able to order WERF publications either through WEF (www.wef.org) or IWAP (www.iwapublishing.com). Visit WERF's website at www.werf.org for details.

IWA Publishing's authorised EU representative for General Product Safety Regulations is Diane D'Arras, 15 rue Duret, 75116 Paris, France, e-mail: safety@iwap.co.uk.

Printed and bound by CPI Group (UK) Ltd, Croydon, CR0 4YY
12/05/2026
02108881-0012